Neutron
Transmutation
Doping of
Semiconductor
Materials

Neutron Transmutation Doping of Semiconductor Materials

Edited by Robert D. Larrabee

National Bureau of Standards
Washington, D.C.

Plenum Press • New York and London

Library of Congress Cataloging in Publication Data

Neutron Transmutation Doping Conference (4th: 1982: National Bureau of Standards)
  Neutron transmutation doping of semiconductor materials.

  "Proceedings of the Fourth Neutron Transmutation Doping Conference held June
1-3, at the National Bureau of Standards, Gaithersburg, Maryland"—T.p. verso.
  Includes bibliographical references and index.
    1. Semiconductor doping, Neutron transmutation—Congresses. I. Larrabee, R.D.
II. Title.
TK7871.85.N48 1982   621.3815′2   83-24522
ISBN-13: 978-1-4612-9675-1   e-ISBN-13: 978-1-4613-2695-3
DOI: 10.1007/978-1-4613-2695-3

Proceedings of the Fourth Neutron Transmutation Doping Conference held, June 1-3, 1982, at
the National Bureau of Standards, Gaithersburg, Maryland

© 1984 Plenum Press, New York
Softcover reprint of the hardcover 1st edition 1984
A Division of Plenum Publishing Corporation
233 Spring Street, New York, N.Y. 10013

DEDICATED TO

JOHN W. CLELAND

These proceedings of the fourth biennial Neutron Transmutation Doping Conference are dedicated to the memory of John W. Cleland of the Oak Ridge National Laboratory. As one of the originators of this series of NTD Conferences, he served as Chairman of the first NTD Conference in 1976 and was an active participant in all subsequent conferences. He will be remembered not only for this participation and his technical contributions, but also for his humorous and informal approach that broke through barriers and promoted open communication with his friends and colleagues. He started preparation of a paper for this conference, but his illness prevented its completion. He died on October 16, 1982.

PREFACE*

This volume contains the papers presented at the Fourth International Neutron Transmutation Doping Conference held at the National Bureau of Standards in Gaithersburg, Maryland on June 1-3, 1982.

Neutron transmutation doping (NTD) is the process of creating desired nonradioactive impurity isotopes from the host atoms of a material by thermal-neutron irradiation and subsequent radioactive decay. This technique is particularly applicable to adding dopant atoms to semiconductors in those cases where it is desired to control the uniformity of doping or where a very small amount of dopant must be added in a controlled fashion. Although the NTD technique has been known and practiced for several decades in the research laboratory, it was not used to any significant extent commercially until the mid-1970s. Since then, the commercial use of NTD silicon has grown rapidly to the point where the 1982 estimated worldwide production rate of NTD silicon was approximately 50 tons per year.

The first formal conference associated with neutron transmutation technology was organized in 1976 by John Cleland and Richard Wood. At this time it had become clear that the NTD process would be implemented on a commercial scale to prepare uniformly doped starting material for the power-device industry.

Two years later, the Second International Conference on Neutron Transmutation Doping in Semiconductors was organized by Jon Meese and held on April 23-28, 1978 at the University of Missouri, Columbia, Missouri. Various aspects of silicon NTD technology were discussed at this conference including irradiation technology, radiation-induced defects, and the use of NTD silicon for device applications. By this time, the interest in NTD silicon had become worldwide and 20% of the conference attendees came from countries outside the U.S.A. The proceedings of this conference were edited by Jon Meese and published by Plenum Press in 1979.

*Contribution of the National Bureau of Standards; not subject to copyright.

The growing use of NTD silicon outside the U.S.A. motivated an interest in having the next NTD conference in Europe. Therefore, the Third International Conference on Neutron Transmutation-Doped Silicon was organized by Jens Guldberg and held in Copenhagen, Denmark on August 27-29, 1980. The papers presented at this conference reviewed the developments which occurred during the two years since the previous conference and included papers on irradiation technology, radiation-induced defects, characterization of NTD silicon, and the use of NTD silicon for device applications. The proceedings of this conference were edited by Jens Guldberg and published by Plenum Press in 1981.

Interest in, and commercial use of, NTD silicon continued to grow after the Third NTD Conference, and research into neutron transmutation doping of nonsilicon semiconductors had begun to accelerate. The Fourth International Transmutation Doping Conference reported in this volume includes invited papers summarizing the present and anticipated future of NTD silicon, the processing and characterization of NTD silicon, and the use of NTD silicon in semiconductor power devices. In addition, four papers were presented on NTD of nonsilicon semiconductors, five papers on irradiation technology, three papers on practical utilization of NTD silicon, four papers on the characterization of NTD silicon, and five papers on neutron damage and annealing. These papers indicate that irradiation technology for NTD silicon and its use by the power-device industry are approaching maturity. However, if the integrated circuit community should require large amounts of uniformly doped silicon of resistivity greater than about 50 $\Omega \cdot$cm, it would create an entirely new demand and foster new research interest in NTD silicon. Because of this new potential large-volume use of NTD silicon and because of the growing interest in NTD of nonsilicon semiconductors, it is very difficult to predict the emphasis of NTD research and development over the next two years. However, that research and development will be reported at the next NTD conference which will be organized by Fritz Vieweg-Gutberlet and held on February 9, 1984 in San Jose, California in conjunction with the ASTM Symposium on Semiconductor Processing.

I would like to thank the members of the 1982 Conference Advisory Committee for their very helpful assistance and advice:

D. Blackburn	National Bureau of Standards
R.S. Carter	National Bureau of Standards
J.W. Cleland	Oak Ridge National Laboratory
E.C. Cohen	National Bureau of Standards
R.F. Fleming	National Bureau of Standards
F. Vieweg-Gutberlet	Wacker-Chemitronic
H. Kramer	Monsanto Corporation
O. Malmros	Topsil A/S

J.M. Meese   University of Missouri
R. Sittig   Brown, Boveri & Co.
J. Washburn   Motorola

We also gratefully acknowledge the financial support received for
this conference from the following organizations:

      National Bureau of Standards
      U.S. Department of Energy
      Monsanto Corporation

A special note of thanks must be given to Jane Walters and J. S.
Halapatz of the National Bureau of Standards for typing portions
of the present manuscript and for their continued guidance and
support in getting the present material prepared for publication,
to the staff of the NBS Semiconductor Technology Program for
their technical consultation in preparing manuscripts for publication, to Ronald Fleming of the NBS Inorganic Analytical Research Division for his careful review of several of the manuscripts, and to my wife, Ramona, for surveying all of the manuscripts for logistical errors.

Robert D. Larrabee, Chairman
1982 NTD Conference

Gaithersburg, Maryland
April 1983

CONTENTS

NEUTRON TRANSMUTATION DOPING OF p-TYPE

CZOCHRALSKI-GROWN GALLIUM ARSENIDE

M. H. Young, A. T. Hunter, R. Baron, O. J. Marsh
and H. V. Winston

Hughes Research Laboratories
3011 Malibu Canyon Road
Malibu, CA 90265

R. R. Hart

Texas A&M University
College Station, TX 77843

ABSTRACT

    We have studied the neutron transmutation doping process in
bulk GaAs grown by the liquid encapsulated Czochralski technique.
By choosing undoped, but initially p-type samples for irradiation,
we observed the effects of transmutation doping with Se and Ge
donors both at low doping levels as added compensation in p-type
samples, and at higher doping levels as added donors in n-type
samples. We found that a measurable fraction of NTD-produced Ge
atoms act as acceptors rather than donors in our material.

    Transmutation dopings of eight different concentrations from
3.8×10^{15} to 6.3×10^{17} (Se + Ge)/cm^3 (determined from nuclear measurements) are included in our study. We characterized irradiated
GaAs samples by temperature dependent resistivity and Hall effect
measurements and measurements of low temperature photoluminescence.
Our p-type starting material has been fully characterized by these
same techniques and found to contain three different acceptor
levels with a total acceptor concentration of 4 to 6×10^{16}/cm^3 and
total initial compensating donor concentration of 1 to 2×10^{15}/cm^3.

    By following individual irradiated samples through a series
of heat treatments, we have studied the effects of radiation

damage on the electrical properties of both p-type and n-type NTD
GaAs samples and observed the annealing of these radiation defects.

INTRODUCTION

Neutron transmutation doping (NTD) of gallium arsenide (GaAs)
is based on the following thermal neutron capture nuclear reactions.

$$^{69}\text{Ga} \ (n,\gamma) \ ^{70}\text{Ga} \xrightarrow[\text{21.1 min.}]{\beta^-} \ ^{70}\text{Ge} \ ,$$

$$^{71}\text{Ga} \ (n,\gamma) \ ^{72}\text{Ga} \xrightarrow[\text{14.1 hrs.}]{\beta^-} \ ^{72}\text{Ge} \ ,$$

$$^{75}\text{As} \ (n,\gamma) \ ^{76}\text{As} \xrightarrow[\text{26.3 hrs.}]{\beta^-} \ ^{76}\text{Se} \ .$$

The relative abundances of the isotopes involved in the reactions
and the cross-sections for these reactions are such that the ratio
of Se and Ge concentrations produced is

$$N_{Se}/N_{Ge} = 1.46.$$

Selenium is a typically shallow substitutional donor in GaAs
with an electronic energy level a few meV from the conduction band
edge (E_c)[1]. Germanium in GaAs is an amphoteric impurity which acts
as a shallow donor (also a few meV from E_c) if situated on a Ga
site and as an acceptor level at $E_V + 0.04$ eV if situated on an As
site[2]. Since, if electronically active, all of the Se atoms and
some portion of the Ge atoms are expected to act as donors, NTD
of GaAs is expected to dope GaAs more n-type.

For this investigation of NTD in GaAs, we chose well-character-
ized, bulk grown p-type samples as our starting material. This
single crystal material was grown in our own laboratory from
pyrolytic boron nitride (PBN) crucibles by the liquid encapsulated
Czochralski (LEC) technique from an arsenic deficient melt with no
intentionally added dopants[3]. This choice of starting material was
advantageous for a number of reasons. Because the material is
conductive rather than semi-insulating, it is readily measurable
and is well characterized before NTD by the experimental techniques
described in the next section. Since we use bulk rather than thin
epitaxial layers, there are no substrate effects to contend with
and no limitations on polishing or etching the samples if surface
effects are suspected. In addition, no intentional, possibly
interfering impurities such as chromium have been added to this GaAs,
and because the material is initially p-type, NTD can convert it
controllably to n-type. One goal in this investigation is to study
the effects of neutron transmutation on the electrical properties

of GaAs and thus to calibrate the donors produced by the transmutation process. We also wish to use this doping technique to assist us in understanding and characterizing impurity or defect electronic levels present in p-type undoped liquid encapsulated Czochralski (LEC) GaAs by adding increments of NTD-produced donors to move the Fermi level in steps through the GaAs bandgap.

The electronic properties of the starting material are schematically presented in Figure 1. Using experimental techniques described in the next section, we have determined that the dominant acceptor levels in this p-type GaAs are located at approximately E_V + 0.07 eV and E_V + 0.2 eV and the concentration associated with each of these levels is $\sim 2 \times 10^{16}/cm^3$. In addition, we measure a residual carbon acceptor concentration of $5-7 \times 10^{15}/cm^3$ and total donor content of $1-2 \times 10^{15}/cm^3$. From other investigations,[4,5,6] we have concluded that the 0.07 eV and 0.2 eV levels are probably due to intrinsic defects related to stoichiometry rather than to impurity atoms. This conclusion results from the observation that these two deeper acceptors are not present in undoped LEC semi-insulating GaAs grown similarly but from an As-rich rather than As-deficient melt.

A schematic explanation of what we might expect to happen as we add donors by NTD to this p-type starting material is shown in Figure 2. The addition of donors moves the Fermi level (E_F) away from the valence band (E_V) toward the conduction band. If a sufficiently high concentration of donors is added, E_F will move to the upper half of the bandgap and the GaAs will be converted to n-type. Our intent in this experiment is to add donors successively, first overcompensating the shallowest acceptors present (carbon) and then one by one overcompensating the deeper acceptor levels. The effect of moving $E_F(T)$ in the material by successively compensating acceptor levels, as illustrated in Figure 2(a)-2(c), would be manifested in Hall effect measurements of hole concentration (p) as a function of temperature, as illustrated in Figure 2(d). The low temperature slope of a plot of ln p versus 1/T corresponds to the shallowest partially compensated acceptor electronic energy level in the p-type samples. This slope is expected to move from 0.026 eV to 0.07 eV to 0.2 eV as the conditions depicted in Figures 2(a), 2(b) and 2(c), respectively, apply.

Our experiment with NTD in GaAs is thus aimed at "titrating" p-type GaAs by adding donors and eventually turning the material n-type. In the process, we can experimentally test our understanding of the electronic structure of the p-type GaAs by comparing our experimental results to the idealized theoretical models presented schematically in Figures 1 and 2. By continuing the NTD process to higher concentrations, we can compare measurements of added electrically active donors with NTD-produced

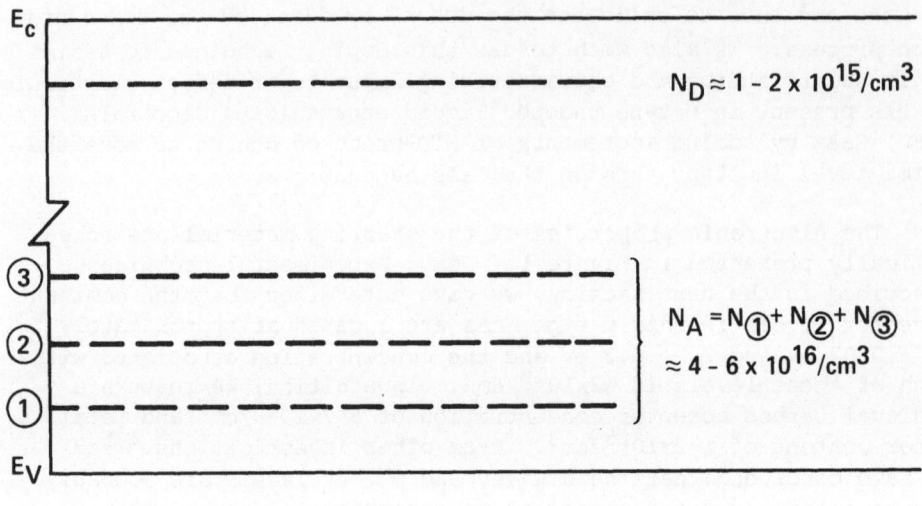

① $E_V + 0.026\ eV$ $N_① \approx 5 - 7 \times 10^{15}/cm^3$ CARBON

②   ~ $E_V + 0.07\ eV$ $N_② \approx 2 - 3 \times 10^{16}/cm^3$ DEFECT LEVEL

③   ~ $E_V + 0.2\ eV$ $N_③ \approx 2 \times 10^{16}/cm^3$ DEFECT LEVEL

Figure 1. LEC p-type GaAs starting material properties

impurity concentrations determined by nuclear measurements of the
transmutation process.

EXPERIMENTAL PROCEDURE

The experimental techniques used in this investigation include
nuclear activity measurements performed in conjunction with the
irradiation process; electrical evaluation by temperature dependent
Hall effect and resistivity measurements; and low temperature
photoluminescence spectral measurements. The three techniques
provide complementary information and, when used together, constitute a powerful means of analyzing the effects of NTD in GaAs.

The GaAs samples selected for transmutation were mounted inside
aluminum containers and irradiated in thermal rotisseries at the
Texas A&M University Research Reactor. Iron standards were included
as neutron fluence monitors. The neutron flux was typically
4×10^{12} n/cm^2-sec, and the cadmium ratio, based on Fe, was about 40.
Irradiation times varied from 2 to 185 hours.

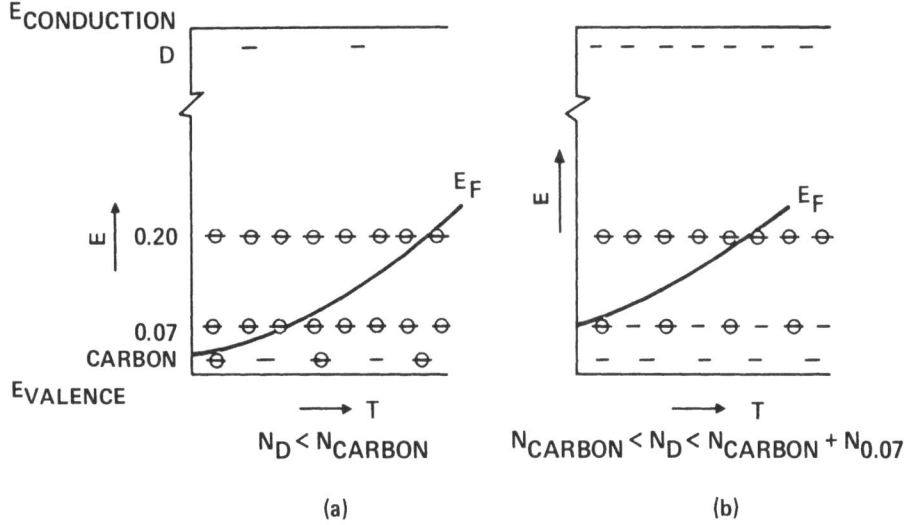

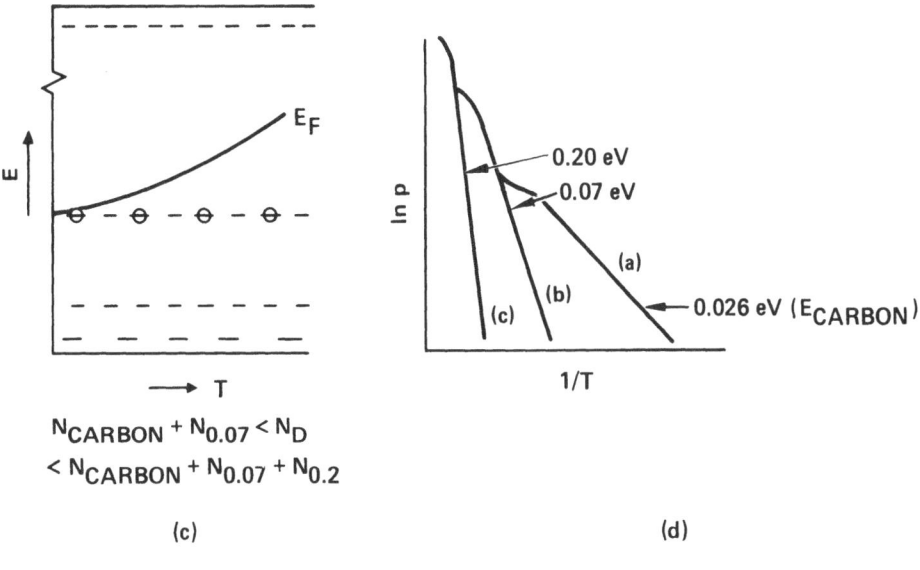

$\ominus$ OCCUPIED, NEUTRAL ⎫

‒ VACANT, CHARGED ⎬ AT T = 0 K
⎭

Figure 2.   Fermi level (E_F) versus T for p-type semiconductor
containing three acceptor levels. Three compensation
cases are shown in (a), (b) and (c). Expected
temperature dependent hole concentration for each of
the cases is illustrated in (d).

Following the procedure discussed in Reference 7, effective
neutron fluxes ϕ_{69}, ϕ_{71}, ϕ_{75} (for ^{69}Ga, ^{71}Ga and ^{75}As reactions,
respectively), and ϕ_{Fe} may be determined from gamma activity
measurements of the GaAs samples and Fe flux monitors. These
effective fluxes allow precise calculations of the concentrations
of transmuted atoms for each of the three NTD reactions, provided
irradiation times are much less than the half lives of the respective radioactive products. This requirement is necessary to account
for possible time variations of the neutron flux during the irradiations. For the longer irradiations, only the determinations of ϕ_{Fe}
satisfy this requirement since the half life of the reaction product,
^{59}Fe, is 44.6 days.

Thus, it is necessary to relate ϕ_{Fe} to the other effective
fluxes. Following a 1-hour irradiation, ϕ_{71}, ϕ_{75} and ϕ_{Fe} were all
determined from gamma activity measurements using a calibrated
Ge(Li) detector. The ratios of ϕ_{71} and ϕ_{75} to ϕ_{Fe} was 1.0 ±15%.
It is assumed that the ratio of ϕ_{69} to ϕ_{Fe} is also 1.0. Since
cross-section data show that only one-third of the Ge atoms produced
are due to neutron absorptions in ^{69}Ga, a 30% error in this estimate
leads to only a 10% error in Ge production.

Therefore, replacing ϕ_{69}, ϕ_{71} and ϕ_{75} by ϕ_{Fe}, the concentrations of Ge and Se, N_{Ge} and N_{Se}, respectively, that were produced
are given by

$$N_{Ge} = \left[\beta_{69} \; \sigma_{69} + \beta_{71} \; \sigma_{71}\right] N \; \Phi_{Fe} \; ,$$

$$N_{Se} = \sigma_{75} N \; \Phi_{Fe} \; ,$$

where

β_{69} and β_{71} = isotopic abundances of ^{69}Ga and ^{71}Ga, respectively,

σ_{69}, σ_{71} and σ_{75} = (n,γ) cross-sections of ^{69}Ga, ^{71}Ga and ^{75}As,
respectively, for neutrons of speed
2200 m/sec,

N = atomic concentrations of Ga and As,

Φ_{Fe} = $\phi_{Fe}t$ = effective neutron fluence (t = irradiation time).

Therefore, on the basis of tabulated abundances and cross-sections,[8]

$$N_{Ge} = 0.065 \; \Phi_{Fe} \; ,$$

$$N_{Se} = 0.095 \; \Phi_{Fe} \; .$$

These equations, together with experimentally determined values of Φ_{Fe} for each irradiation, were used to calculate N_{Ge} and N_{Se}. Including uncertainties in effective flux ratios and cross-sections, as well as experimental errors in the Φ_{Fe} measurements, we estimate the total error in calculated Ge and Se concentrations to be ±20%.

Analysis of Hall effect data as a function of temperature provides a means of measuring the donor content in our GaAs samples. We are thus able to compare electrically active added donor content to the NTD-produced impurity concentrations determined from nuclear measurements. The Hall effect analysis also allows us to determine concentrations and energy levels (E) of impurities or defects in the p-type GaAs samples if the Fermi level in the material moves near E at some temperature over the range of measurements. The technique thus provides a means of identifying and measuring undercompensated acceptor content in the samples.

The low temperature photoluminescence technique used here measures donor-to-acceptor or conduction-band-to-acceptor lumin-escence. It provides an accurate determination of the position of acceptor electronic levels in the GaAs, permitting positive identi-fication of impurities or defects with known luminescence lines. Identification of lines due to specific impurities or defects can be made using luminescence techniques regardless of the position of the Fermi level in the material. Little detailed information concern-ing an acceptor level can be obtained from Hall effect if that acceptor is overcompensated. However, the presence of specific acceptors can be detected by luminescence techniques even in n-type samples. On the other hand, our luminescence data do not provide the quantita-tive information obtainable from Hall effect measurements.

Individual samples, cut from one characterized wafer, were neutron irradiated at eight different dose levels. The added Se and Ge impurity content, as determined from our nuclear activity measurements and analysis, are tabulated in Table 1. Following irradiation of the samples, we applied suitable contacts for electrical measurements and provided SiO_2 capping to protect the surfaces. Samples doped at each of the eight fluence levels presented in Table 1 received anneals at 830°C/20 min. This anneal procedure is patterned after standard processing steps for fully annealing Se and Si implanted layers in GaAs for IC device applications.[9] Two other samples - one doped with the lowest NTD dose (3.8×10^{15}/cm^3) and one with the highest dose (6.3×10^{17}/cm3) - were characterized following each of a sequence of anneals from room temperature to 830°C. Control samples which received no irradiation were processed identically to the NTD samples to check for changes produced in the material by the anneal step.

Table 1.   Transmutation Doped Samples

$(N_{Ge} + N_{Se})/cm^3$ Added to Samples	N_{Se}	N_{Ge}
3.8×10^{15}	2.3×10^{15}	1.5×10^{15}
8.5×10^{15}	5.0×10^{15}	3.5×10^{15}
1.7×10^{16}	1.0×10^{16}	6.8×10^{15}
2.7×10^{16}	1.6×10^{16}	1.1×10^{16}
7.0×10^{16}	4.1×10^{16}	2.9×10^{16}
1.5×10^{17}	8.8×10^{16}	5.9×10^{16}
2.8×10^{17}	1.7×10^{17}	1.1×10^{17}
6.3×10^{17}	3.7×10^{17}	2.6×10^{17}

EXPERIMENTAL RESULTS AND DISCUSSION

Electrical Measurement

The results of room temperature measurement of the electrical properties of our group of eight annealed NTD GaAs samples are summarized in Table 2. The total NTD dose ($N_{Se} + N_{Ge}$), the carrier concentration and carrier type (negative values of concentration indicate n-type), along with carrier mobility at room temperature are indicated in Table 2. Note that following an NTD dose sufficient to produce 7×10^{16} atoms/cm^3, the material is converted to n-type. Our characterization of the starting material indicates a total of $4-6 \times 10^{16}$ acceptors/cm^3 (see Figure 1) initially present in the samples. Therefore, 7×10^{16} donors/cm^3 would indeed be expected to just overcompensate the p-type material. The results presented in Table 2 show that the p-type samples become progressively less p-type and the n-type samples progressively more n-type with increasing NTD dose.

Hall effect measurements of the temperature-dependent hole concentration for NTD p-type samples, along with control samples, are presented in Figure 3. The plots for the control samples 23 and 24 and the NTD samples 1 and 2E are in accordance with the case of Figure 2(a) because they exhibit low temperature slopes of log p vs. 1/T, corresponding to the energy level of carbon. Sample 3 is a transition case in which the low temperature slope is still determined mostly by the carbon energy.   In the low

Table 2. Room Temperature Results for Hall Effect Samples
Annealed 830°C/20 Min

Sample No.	NTD Dose/cm^3	n(−) or p(+) in cm^{-3}	μ, cm^2/V-sec
1, 2	3.8×10^{15}	$+ 2.3 \times 10^{16}$	360
3	8.5×10^{15}	$+ 2.4 \times 10^{16}$	341
4	1.7×10^{16}	$+ 1.9 \times 10^{16}$	337
10	2.7×10^{16}	$+ 8.6 \times 10^{15}$	242
12	7×10^{16}	$- 1.6 \times 10^{16}$	1251
20	1.5×10^{17}	$- 7.7 \times 10^{16}$	3960
15	2.8×10^{17}	$- 2.3 \times 10^{17}$	3631
16, 18	6.3×10^{17}	$- 4.9 \times 10^{17}$	3110

temperature regime where p varies as $\exp(-\Delta E_{carbon}/KT)$, the hole
concentration is given by[10]

$$p \sim \frac{N_{carbon} - N_D}{N_D} \quad \frac{N_v}{g} \quad \exp(-E_{carbon}/kT)$$

$N_V \equiv$ effective valence band density of states

$g \equiv$ degeneracy associated with the carbon acceptor

$N_D \equiv$ donor concentration

The measurements of hole concentrations for the control samples
(No. 23 and No. 24), sample 1 and sample 2E, and sample 3 are
consistent with an increasing donor concentration as NTD dose is
increased. The effect of varying N_D is essentially to shift the
log p vs. 1/T plot vertically.

The 0.04 eV slope observed for sample 4 implies that a new
acceptor level has been introduced between the 0.026 eV level of
carbon and the 0.07 eV level depicted in Figures 1 and 2. A detailed
analysis of the data for sample 2E shown in Figure 3 (following the
procedure described in references 11 and 12) also indicates the
appearance of a new acceptor level at $\sim E_V + 0.04$ eV. In this semi-
log plot of hole concentration vs. 1000/T, the carbon acceptor

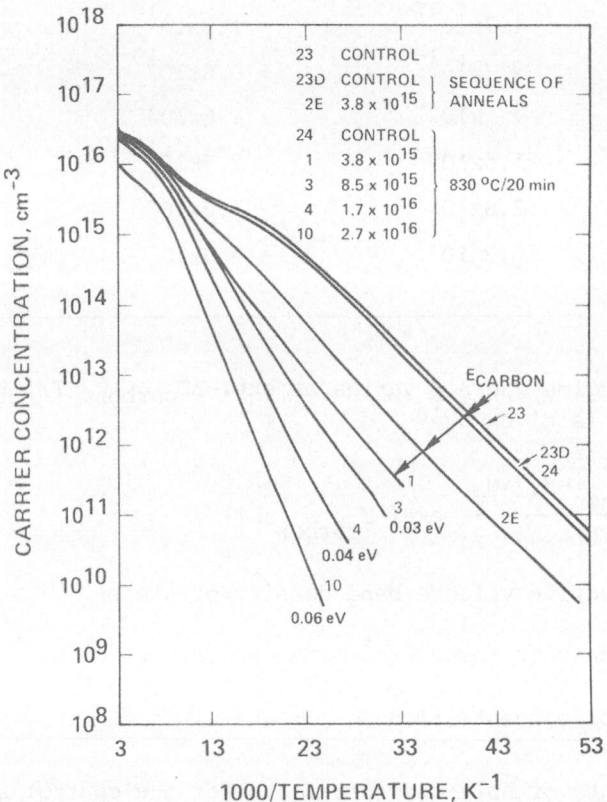

Figure 3.   Hall effect data for p-type NTD samples

activation energy appears at the lowest temperatures for sample 2E, but in the intermediate temperature range (18 to 33 in 1000/T), the activation energy of the Ge acceptor ($\sim$0.04 eV) dominates the measured slope. Photoluminescence measurements on these samples (described below) confirm the presence of a new level and positively identify it as Ge on an acceptor site.

The experimental results for sample 10 shown in Figure 3 imply that the case depicted in Figure 2(b) applies at this stage of neutron transmutation doping; enough donors have been added by NTD so that N_D exceeds N_{carbon} but is smaller than $N_{carbon} + N_{0.07}$.

The measured free electron concentration for the four n-type samples is shown in Figure 4. Because the donor levels in GaAs are very shallow, they remain fully ionized in the temperature range of our experiment, so that the measured electron concentration is practically temperature independent. This measured n for each sample is approximately equal to total donor minus total acceptor concentration.

Figure 5 shows the measured added electrically active donor concentration in our eight NTD samples as a function of N_{Se} and of ($N_{Se} + N_{Ge}$) added by transmutation as determined from nuclear activity measurements. The uncertainty in determining added donor content in the p-type samples is large because of the complexity of analyzing material with multiple independent acceptor levels in closely compensated cases. The added donors can be much more accurately determined in the more highly doped n-type samples. Setting the Hall r-factor (ratio of Hall mobility to drift mobility) equal to unity introduces an uncertainty of the order of $\sim$20% for the n-type samples. Estimated uncertainties in the nuclear activity measurements of N_{Se} and N_{Ge} were discussed above. The results shown in Figure 5 imply that all of the selenium and a substantial fraction of the Ge atoms introduced by transmutation act as donors following the 830°C/20 min anneal. However, from low temperature photoluminescence measurements presented below, we know that a fraction of Ge atoms produced by transmutation are on acceptor rather than donor sites in these samples.

Hall analysis also indicates no substantial change in electrical properties of control samples due to annealing procedures. We also observe no major changes in the concentrations of the dominant acceptors (0.07 eV and 0.2 eV defects) in the p-type NTD samples for which these levels can be analyzed.

Low Temperature Photoluminescence

Figures 6 and 7 show relative luminescence spectra for the four p-type and four n-type samples, respectively. The spectral

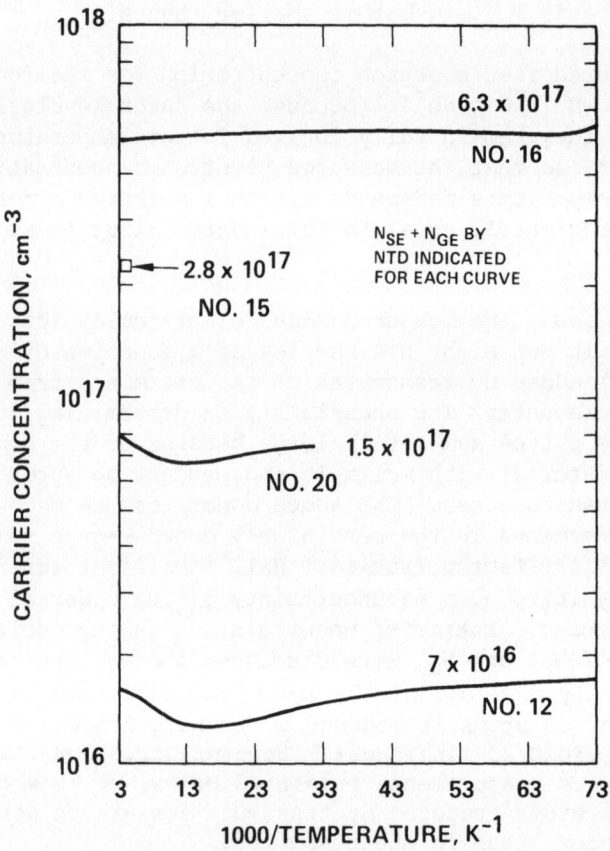

Figure 4. Hall effect data for n-type NTD samples

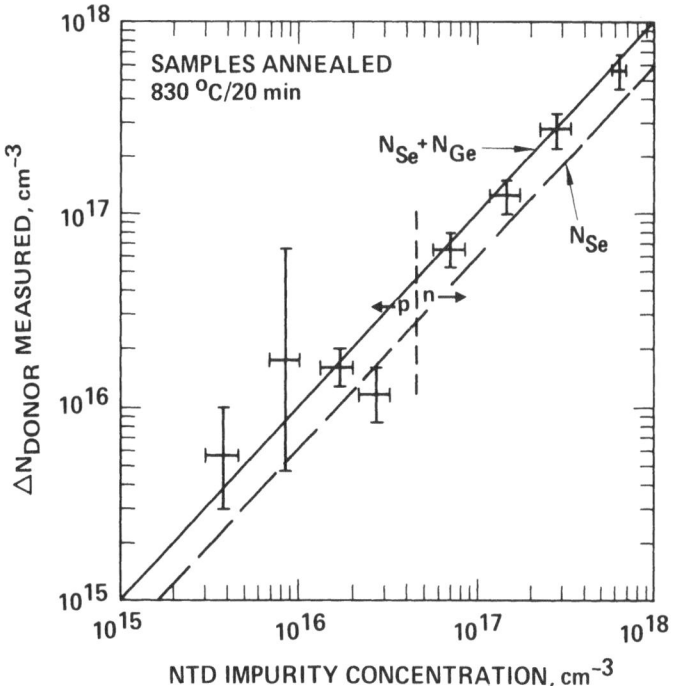

Figure 5.   Measured added donors vs NTD produced
            impurity content

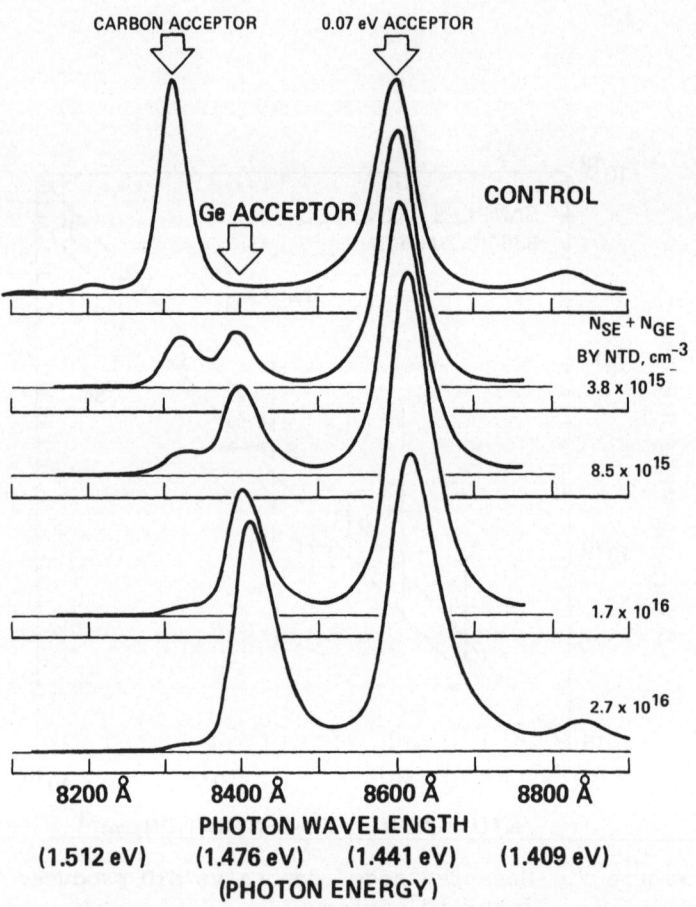

Figure 6.   Photoluminescence spectra for p-type NTD samples;
displayed with intensity of luminescence due to .07 eV
acceptor held constant for the four spectra.

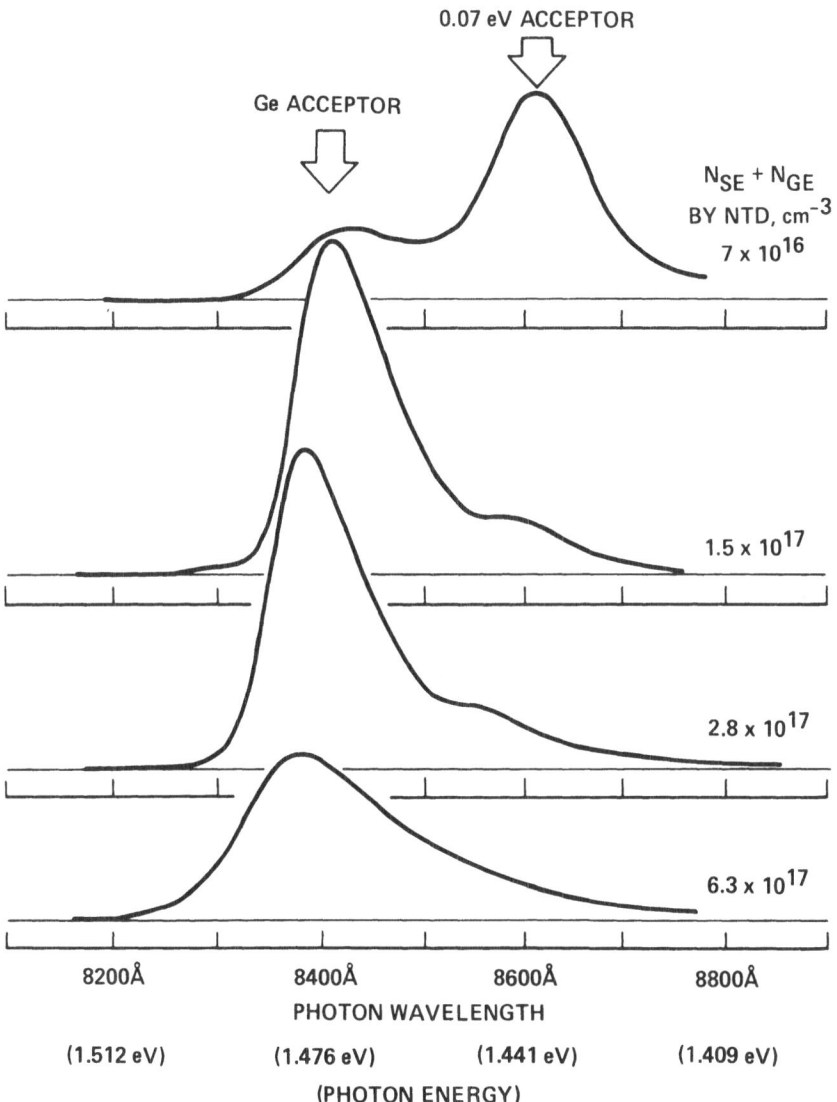

Figure 7. Relative photoluminescence spectra for four n-type NTD samples. The four spectra are not normalized with respect to each other.

positions indicated by arrows for carbon acceptor, the Ge acceptor, and 0.07 eV acceptor correspond to donor (or band) to acceptor luminescence lines. The most important conclusion to be drawn from a comparison of the spectra for the control and eight NTD samples is that Ge acceptors not present in the "starting material" control sample are introduced by the NTD process. The increase in intensity of the Ge acceptor line with increasing dose relative to both the carbon and 0.07 eV acceptor lines indicates that Ge acceptor content increases with increasing transmutation doping. Therefore, some of the Ge atoms produced by NTD in these samples are acting as acceptors rather than donors.

We have found that the relative intensities of the carbon acceptor and 0.07 eV acceptor luminescence lines are laser pump power dependent. No firm conclusions about possible changes in carbon acceptor concentration can be drawn from the luminescence data taken to date.

Photoluminescence measurement studies of the control and eight annealed NTD samples at longer wavelengths indicate another new line present only in NTD samples at about 9450 Å. The intensity of this line increases with increasing NTD dose. This result is preliminary but indicates there may be deep NTD-related centers present in the material following the 830°C/20 min anneal used here.

Anneal Study

Sample 18, which received 6.3×10^{17} NTD atoms/cm^3, and No. 2, which received the lightest dose of 3.8×10^{15} NTD atoms/cm^3, together with a control sample (No. 23), were followed through a sequence of one-hour anneals at six different temperatures from room temperature to 830°C.

After irradiation of No. 18 (6.3×10^{17} NTD/cm^3), no reliable Hall effect measurements could be made until after a 660°C anneal, but a room temperature van der Pauw resistivity measurement on the sample indicates the material to be 2×10^4 Ω-cm following irradiation and before any elevated temperature annealing. Figure 8 shows the measured electrical properties of sample No. 18 following higher temperature anneals. The samples reached a final net n-type doping level following a 740°C anneal. The change in measured donor minus acceptor content between 660°C and 740°C anneals may indicate either an increase in electrically active donors or decrease in compensation (possible due to radiation damage) in the sample.

The measured temperature dependent hole concentration for lightly neutron transmutation doped p-type sample No. 2 following each sequential anneal is displayed in Figure 9. After irradiation and before any elevated temperature annealing, the electrical properties of sample No. 2 are dominated by deep electronic energy levels located

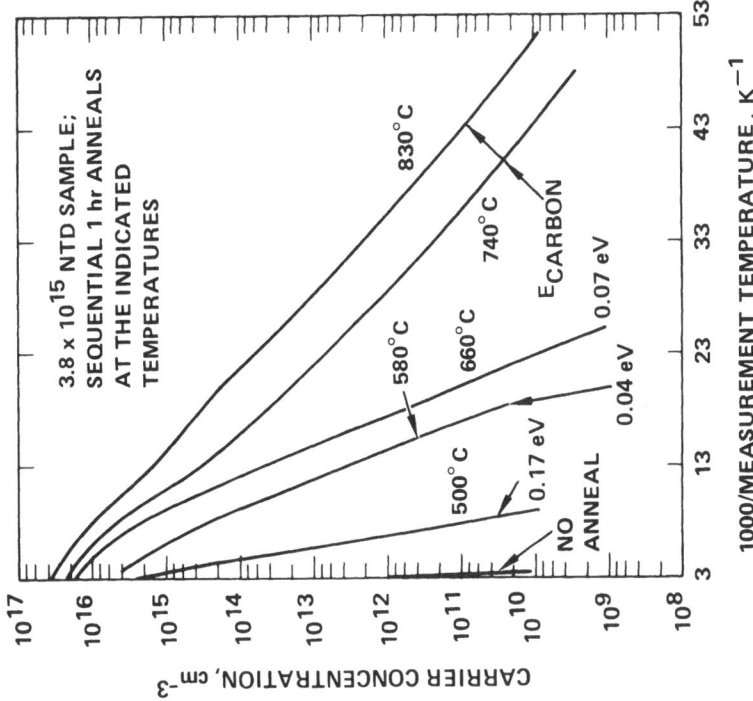

Figure 9. Temperature dependent hole concentration as a function of anneal temperature for sample No. 2 determined by Hall effect measurements.

Figure 8. Electrical properties of n-type GaAs NTD sample vs anneal temperature.

at approximately E_v + 0.5 eV and probably due to radiation damage. The principal result of progressively higher-temperature annealing of this sample is to remove compensating donors resulting, most likely, from residual radiation damage. The results of methodically reducing donor content in this p-type sample by annealing are comparable to but reversed in effect to that of increasing compensation by NTD as depicted in Figure 2. Following the 500°C anneal, the observed low temperature slope of the log p versus 1/T plot in Figure 9 is ~0.2 eV - the expected result if the case depicted in Figure 2(c) applies. After a sufficient donor concentration is removed to undercompensate the 0.07 eV acceptor defect level (corresponding to the case shown in Figure 2(b)), a low temperature slope of 0.07 eV is expected and is indeed observed following anneals at 580°C and 660°C. The case depicted in Figure 2(a) applies once the donor content has been reduced sufficiently that N_{carbon} exceeds N_D. Following the 740°C and 830°C anneals shown in Figure 9, the carbon acceptor activation energy appears at low temperatures in accordance with what is expected for that case.

Our annealing studies clearly show that the electrical properties of NTD GaAs depend sensitively on anneal history. The p-type sample No. 2 followed through the series of anneals shown in Figure 9 never reached a "final" state. Also, even though sample No. 1 shown in Figure 3 and sample No. 2 received the same NTD dose (3.8×10^{15}/cm^3) and were annealed at the same temperature (830°C), the low temperature electrical properties of the two are substantially different (compare sample No. 2E and sample No. 1 in Figure 3). The anneal history for the two is different. Sample No. 1 received a single 830°C/20 min anneal, while sample No. 2 received a sequence of anneals ending with an 830°C/1 hour anneal. Likewise, n-type samples No. 16 (see Figure 4) and No. 18 (see Figure 8) received the same NTD dose (6.3×10^{17}/cm^3) but the room temperature carrier concentration for No. 18 is ~4.2×10^{17}/cm^3 following the sequence ending with 830°/1 hour, while the measured electron concentration for No. 16 is ~4.9×10^{17}/cm^3 following a single 830°C/20 min anneal. This difference is relatively small but is still outside estimated systematic experimental measurement errors.

SUMMARY AND CONCLUSIONS

We neutron-transmutation-doped p-type LEC-grown bulk GaAs samples containing 4-6×10^{16} acceptors/cm^3 "as-grown" with total Se plus Ge concentrations ranging from 3.8×10^{15} to 6.3×10^{17}/cm^3. We investigated the electronic properties of the samples following various anneal procedures using temperature dependent measurements of Hall effect and resistivity and low temperature photoluminescence spectral measurements. Our analysis of the four more lightly

irradiated samples, which remained p-type, yielded results in accord
with the multiple acceptor level model we have used to describe the
p-type starting material. Added donor concentrations measured
following suitable annealing to remove radiation damage agree roughly
with the concentrations of Se + Ge added by transmutation as determined by nuclear activity measurements. However, some fraction of
the Ge atoms introduced by transmutation end up on acceptor rather
than donor sites. We expect that the proportion of Ge acceptors
produced might be dependent on processing procedures, anneal
history, or possibly the stoichiometry of the GaAs starting material.
A similar NTD study of material such as semi-insulating LEC GaAs
from an As-rich rather than As-deficient melt is needed to test for
stoichiometry-related effects.

No substantial change was observed in any control samples due
to process or anneal procedures. Concentrations of the dominant
acceptors present in the starting material were not observed to be
altered appreciably by the NTD process. Some evidence of deep
NTD-related electronic levels appear in photoluminiscence spectra
following an elevated temperature anneal.

REFERENCES

1. S. M. Sze, Physics of Semiconductor Devices (Wiley-Interscience,
   New York, 1969), p. 30.
2. D. J. Ashen, P. J. Dean, D. T. Hurle, J. B. Mullin, A. M.
   White, P. D. Greene, J. Phys. Chem. Solids 36, 1041 (1975).
3. H. Winston, H. Kimura, A. T. Hunter, C. B. Afable, R. Baron,
   J. P. Baukus, M. H. Young and O. J. Marsh, "Growth and
   Characterization of LEC-PBN Undoped Gallium Arsenide," paper
   presented at AACG/West Sixth Conference on Crystal Growth,
   June 1-4, 1982, Fallen Leaf Lake, California.
4. A. T. Hunter, R. Baron, J. P. Baukus, H. Kimura, M. H. Young,
   H. Winston and O. J. Marsh, "Temperature-Dependent Hall
   Measurements on LEC GaAs," paper presented at 2nd
   Conference on Semi-Insulating III-V Materials, Evian,
   France, April 19-21, 1982.
5. P. O. Yu and D. C. Reynolds, J. Appl. Phys. 53, 1263 (1982).
6. K. R. Elliot, D. E. Holmes, R. T. Chen and C. G. Kirkpatrick,
   Appl. Phys. Lett. 40, 898 (1982).
7. R. R. Hart, L. D. Albert, N. G. Skinner, M. H. Young. R. Baron
   and O. J. Marsh, in Neutron Transmutation Doping in
   Semiconductors, Jon M. Meese, Ed. (Plenum, New York, 1979),
   p. 345.
8. S. F. Mughabghab and D. I. Garber, BNL-325 (1973).
9. C. Lawrence Anderson, "Material-Process Interactions in the
   Annealing of Gallium Arsenide" in Laser and Electron-Beam
   Interactions with Solids, edited by B. R. Appleton and
   G. K. Celler (Elsevier Science Publishing Company, 1982),
   p. 653.

10. J. S. Blakemore, <u>Semiconductor</u> <u>Statistics</u> (Pergamon, Oxford, 1962), Chap. 3.
11. M. H. Young, Doctoral Thesis, University of California at Los Angeles, 1980.
12. M. H. Young, O. J. Marsh and R. Baron, J. Appl. Phys. <u>50</u>, 3755 (1979).

NTD GERMANIUM: A NOVEL MATERIAL FOR LOW TEMPERATURE BOLOMETERS*

E.E. Haller, N.P. Palaio and M. Rodder,

Dept. of Materials Science and Mineral Engineering
and Lawrence Berkeley Laboratory
University of California
Berkeley, California 94720 U.S.A.

W.L. Hansen

Lawrence Berkeley Laboratory
University of California
Berkeley, California U.S.A. 94720

E. Kreysa

Max-Planck-Institute for Radioastronomy
Bonn, FRG

ABSTRACT

Six samples of ultra-pure ($|N_A - N_D| \leq 10^{11}$ cm^{-3}), single-crystal germanium have been neutron transmutation doped to p-type with neutron doses between 7.5×10^{16} and 1.88×10^{18} cm^{-2}. After thermal annealing at 400°C for six hours in a pure argon atmosphere, the samples were characterized with Hall effect and resistivity measurements between 300 and 0.3 K. The results show that the resistivity in the low temperature, hopping conduction regime can be approximated by $\rho = \rho_0 \exp(\Delta/T)$. The three more heavily-doped samples show values for ρ_0 and Δ ranging from 430 to 3.3 Ω cm and

---

*This work was supported in part by NASA Contract No. W-14,606 under Interagency Agreement with the Director's Office of Energy Research, Office of Health and Environmental Research, U.S. Department of Energy under Contract No. DE-AC03-76SF00098.

from 4.9 to 2.8 K respectively. The excellent reproducibility of
neutron transmutation doping and the values of ρ_0 and Δ make NTD
germanium a prime candidate for the fabrication of low-temperature, low-noise bolometers. The large variation in the tabulated
values of the thermal neutron cross sections for the different germanium isotopes makes it clear that accurate measurements of these
cross sections for well defined neutron energy spectra would be
highly desirable.

INTRODUCTION

   Low-temperature semiconductor bolometers have gained in importance as very long wavelength photon power sensors in recent
years.[1] The possibility of performing infrared astronomy outside
the atmosphere from satellites, space probes and the space shuttle
has opened the medium and long infrared regions of the electromagnetic spectrum. Until recently, the vast amount of information
contained in this part of the spectrum has not been actively
sought by man.   This is in part because the atmosphere does not
transmit these longer wavelengths, but also because man's vision
has naturally given great emphasis to the visible light range of
the electromagnetic spectrum.

   The potential use of doped semiconductors as photon power
detectors was recognized some forty years ago at Bell Laboratories. The photon absorption in semiconductor bolometer material
raises the temperature which, in turn, leads to a resistivity
change, a parameter which can be measured electrically. The first
liquid helium temperature germanium bolometers were designed and
operated by Low.[2] These cryogenic bolometers have found a wide
range of applications because they are sensitive and rugged. One
major problem however, has remained unsolved--the reliable fabrication of highly-doped and highly-compensated semiconductor single-
crystal bolometer material. This problem becomes more important
with decreasing bolometer operating temperature. As we shall show
that, the interest in going to ever lower operating temperatures
persists because of signal-to-noise considerations. The small
doping fluctuations which one encounters in all melt-doped and
grown crystals[3] becomes important in the low-temperature conduction regime. At temperatures below 1 K, dopant concentration
fluctuations of a few percent lead to resistivity fluctuations of
more than an order of magnitude. This strong detrimental effect
has stimulated the search for improved doping techniques. Ion
implantation together with planar technology have been used to fabricate thin silicon bolometers.[4] Though some outstanding bolometers have been fabricated with this technology, one would like
to obtain homogenously-doped semiconductor single crystals in quantities which allow the fabrication of large numbers of identical
bolometers. Further disadvantages of the ion implantation approach

are the high level of technology and the extreme control of surface cleanliness which are required. The large number of high temperature processing steps together with the delicate structure of these bolometers are factors which may be responsible for the low yield of useful devices obtained with this technology. In contrast, neutron transmutation doping of germanium yields bulk-doped single crystal material. A few low temperature processing steps yield large numbers of identical bolometers.

The low-temperature resistivity dependence of neutron-transmutation-doped germanium has been explored by Fritzsche[5] over twenty years ago but surprisingly, nobody has published the application of this material to low temperature bolometer fabrication.

SEMICONDUCTOR BOLOMETERS

The thermal equivalent circuit of a low-temperature bolometer is illustrated in Fig. 1. The conservation of energy requires that the photon power which is absorbed in the bolometer $\eta \bar{P}$ be equal to the heat flowing through the thermal link of thermal conductance G (typically a very fine wire) and the heat required to change the temperature of the bolometer $H(d\theta/dt)$ be given by:

$$\eta \bar{P} = H(d\theta/dt) + G\theta \qquad (1)$$

The temperature of the bolometer in the dark is T and θ is the bolometer temperature increase under illumination. An additional term which can be neglected at cryogenic temperatures is the energy dissipated by the bolometer through radiation. Typically the incident power consists of a constant background part P_0

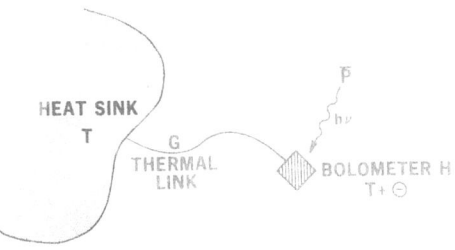

Fig. 1. Schematic of a typical bolometer thermal equivalent circuit.

and a time dependent signal part $P_S = A \exp(i\omega t)$. Substitution
into Eq. (1) and solving for Θ_S yields:

$$\Theta_S = \eta A \, (G^2 + \omega^2 H^2)^{-\frac{1}{2}} \, . \, \exp(i\omega t + \psi) \qquad (2)$$

with the phase shift:

$$\psi = \arctan (\omega H/G) \qquad (3)$$

In order to maximize Θ_S for a given signal amplitude A, one wants
to minimize G and ωH. The thermal response time constant τ can be
defined as:

$$\tau = \omega^{-1} \qquad (4)$$

at the radian frequency where $G = \omega H$. It follows that:

$$\tau = H/G \qquad (6)$$

Equation 6 shows that H should be made as small as possible, cer-
tainly much smaller than G in order to get short response times
and the highest possible temperature fluctuation Θ_ω. The electrical circuit for the measurement of the resistivity fluctuations
caused by the temperature changes is shown in Fig. 2. A constant
current I leads to a voltage V across the bolometer. The signal
voltage V_S is caused by resistance changes δR. The DC component
of the voltage drop across the bolometer is separated from the
time dependent signal part by the capacitor C. Using the conventional definition of the temperature coefficient α:

$$\alpha = \frac{1}{R} \, (dR/dT) \qquad (7)$$

we find for V_S:

$$V_S = I \quad \alpha \quad R \quad \Theta_S$$
$$\quad = I \, \alpha \, R \, \eta \, P_S \, (G^2 + \omega^2 H^2)^{-\frac{1}{2}} \qquad (8)$$

The responsivity RESP in V/W, a figure of merit for a bolometer is
given by:

$$RESP = V_S/P_S = \eta I \alpha R (G^2 + \omega^2 H^2)^{-\frac{1}{2}} \qquad (9)$$

Contrary to the simple interpretation of Eq. (9), one obviously

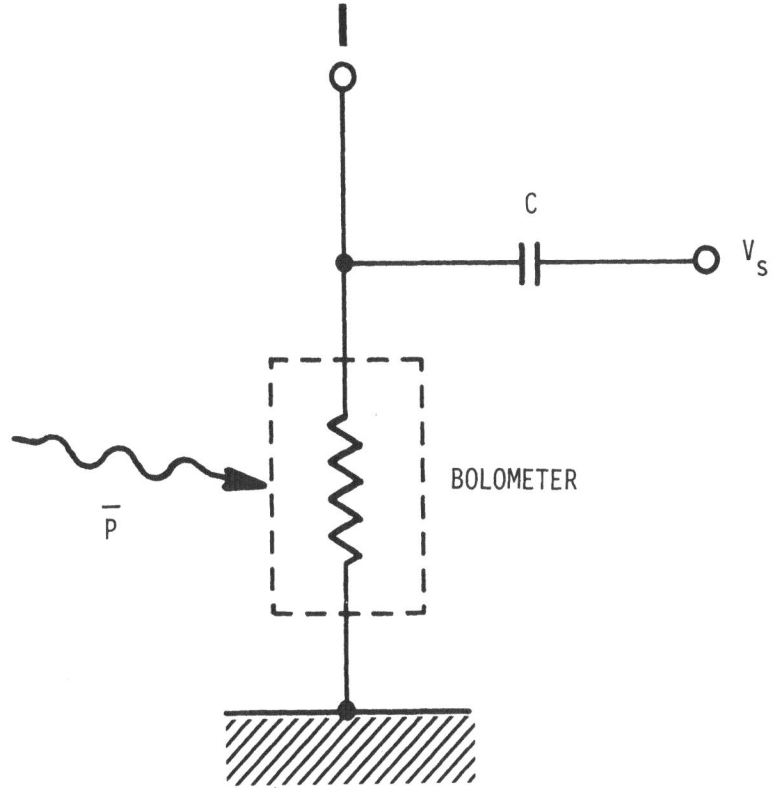

Fig. 2.   Electrical circuit for bolometer operation.

cannot choose I and R to have arbitrarily large values and thereby obtain arbitrarily large responsivities. The current through the bolometer leads to an internal source of power and a corresponding increase in bolometer operating temperature:

$$P_{int.} = I^2 R \qquad\qquad\qquad (10)$$

The proper current I is chosen so that the responsivity is a maximum. Furthermore, a careful analysis of all the noise sources discloses that the optimum electrical amplifiers show the lowest excess noise when they operate with a source input resistance of 10^5 to $10^7 \Omega$. It is not the aim of this paper to review all the aspects of the complicated and interrelated noise terms for a given bolometer. Such analyses have been published recently.[6] The foregoing equations show that a large value of α, a small thermal mass H and an optimum resistance between 10^5 and $10^7 \Omega$ are the basic requirements for a high-performance low-temperature semiconductor bolometer.

It should also be mentioned that the ideal bolometer approaches
a fundamental noise limit, the fluctuations which are inherently
present in any photon stream. In the Poisson statistics approximation one can describe the fluctuation ΔN in the photonstream $\bar{N}$ by:

$$\Delta N = \bar{N}^{\frac{1}{2}} \tag{11}$$

with

$$\bar{N} = \bar{P}/h\nu \tag{12}$$

and for a detection system of band width B and of quantum efficiency η, we obtain:

$$\Delta P = \left(h\nu\bar{P}B/\eta \right)^{\frac{1}{2}} \tag{13}$$

Equation 13 describes the photon fluctuations in a photonstream in
terms of power fluctuations and is a fundamental limit that cannot
be circumvented.

ELECTRICAL CONDUCTION MECHANISMS IN SEMICONDUCTORS

Four mechanisms which lead to electrical conduction in a semiconductor crystal are schematically illustrated in Fig. 3. Mechanism 1 shows the thermal generation of electrons and holes across
the bandgap. In silicon and germanium this term becomes negligible
at low temperatures where $kT \ll E_{gap}$. Mechanism 2 illustrates the
generation of free charge carriers (in this particular example--
electrons) by ionization of shallow donors. This mechanism can
also be neglected at very low temperatures because $kT \ll E_C - E_D$.
Mechanisms 3 ("banding") and 4 ("hopping") occur in semiconductors
which are heavily doped, and heavily doped and compensated respectively. The charge carrier moves from one impurity to the next
without reaching the band. Mechanism 3 is characterized by an impurity concentration which is high enough to lead to substantial
overlap between neighboring wave functions. The individual impurity states form an impurity band. The average interimpurity distance and its fluctuations are the critical parameters. The transition at very low temperatures from the insulator regime at large
interimpurity distances to the metallic conduction regime at small
interimpurity distances occurs abruptly. It is called the metal-
insulator transition[7,8] or somewhat incorrectly, Mott-transition.
Hopping conduction (mechanism 4) occurs when compensating or minority impurities create a number of majority impurities which remain
ionized down to absolute zero. In this case, charge carriers can
"hop" from an occupied majority impurity site to an empty site.
It is this mechanism which has been proven to be useful for low-
temperature bolometer applications. As in the case of "banding"
conduction, the interimpurity distance and its fluctuations are of

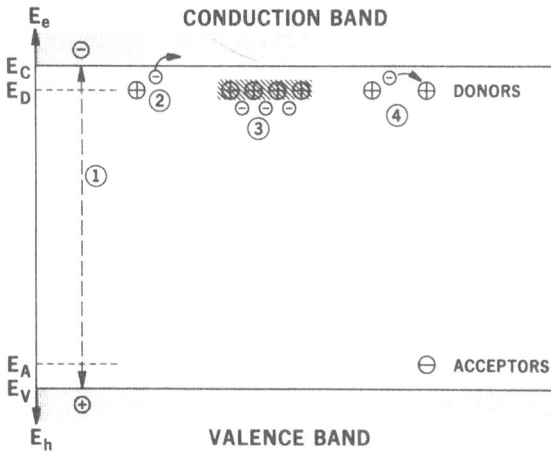

Fig. 3. Electrical conduction mechanisms in semiconductors.

critical importance. It is almost impossible to obtain large crystals with sufficiently good impurity distribution homogeneity by
conventional melt doping techniques. Trial and error have been the
major approach to solving this problem. Extensive and costly testing of many single crystal samples has yielded some useful material
which is carefully guarded by the owner. In view of the fact that
Fritzsche and coworkers have used neutron-transmutation-doped germanium to study the fundamental conduction mechanisms in semiconductors[5] already in the early 60's , it is surprising that to our
knowledge, there exists no record in archival journals of the use
of NTD germanium for low-temperature bolometer fabrication.

NEUTRON TRANSMUTATION DOPING OF ULTRA-PURE GERMANIUM

     The advantages of neutron transmutation doping over conven-
tional doping techniques have been discussed in great detail.[9,10]
For the specific application of NTD germanium discussed in this
paper, the basic advantage of extreme homogeneity in the distributions of the shallow acceptors and compensating donors becomes
more important than in typical NTD silicon applications. To illus-
trate this point it may be useful to mention that near the onset of

"hopping" conduction, an increase in doping concentration of slightly more than one order of magnitude leads to resistivity changes at T = 4 K by seven orders of magnitude. This shows that not only the homogeneity but also the net-dopant concentration itself is extremely critical.

The transmutation of the stable germanium isotopes via the capture of thermal neutrons is well understood. Table 1 contains all the information relevant to NTD of germanium. The listed values are well known with the exception of the thermal neutron capture cross sections. We quote the values of σ_n of three different sources.[10,11,12]

The information in Table 1 permits the computation of the acceptor and donor concentrations for a known neutron exposure. Not only are these concentrations important but the ratio of the sum of all minority dopants (donors) and the sum of all majority dopants (acceptors) i.e., the compensation K, is crucial for the low temperature conduction. For the case of germanium, one obtains K from the following equation:

$$K = (\Sigma \text{ donors/cm}^{-3})/(\Sigma \text{ acceptors/cm}^{-3})$$

$$= (N_{As} + 2N_{Se})/N_{Ga} \tag{14}$$

The substitutional selenium impurities are double donors providing two electrons for compensation. Therefore they are counted twice in the sum of donors. Using the different values for σ_n, one finds K ranging from 0.322 to 0.405 for crystals with negligible initial donor and acceptor concentrations. It would be of great help for both the basic understanding of the hopping conduction as well as for the application of neutron-transmutation-doped germanium as bolometer material, if these cross sections could be accurately evaluated in one or more well characterized nuclear reactors.

In order to obtain the above K values and thus take full advantage of neutron transmutation doping, one should choose the purest available crystals as a starting material. Germanium is, in this respect, ideally suited for NTD because it can be purified today down to concentrations of $\leq 10^{11}$ cm^{-3}.[14] Such low concentrations are negligible when compared with the dopant concentrations after NTD in the low 10^{16} cm^{-3} range. The concentrations of electrically inactive impurities such as hydrogen, carbon, oxygen and silicon can be as high as 10^{14} cm^{-3}. Of all the isotopes of these impurities only $^{30}_{14}$Si transmutes to an electrically active impurity, phosphorus, a shallow donor. With only one silicon atom in every 4.4 x 10^8 germanium atoms and only 3% of all silicon atoms being $^{30}_{14}$Si which has a neutron capture cross section much smaller than the germanium isotope cross sections, we can estimate that less than one

TABLE 1.

Isotope	Abundance (%)	Neutron Capture Cross Sections (barn)			Neutron Capture and Decay Reactions	Dopant Type
		(11)	(12)	(13)		
$^{70}_{32}\text{Ge}$	20.5	3.4	3.2	3.25	$^{70}_{32}\text{Ge}(n,\gamma)^{71}_{32}\text{Ge} \xrightarrow[11.2d]{EC} {}^{71}_{31}\text{Ga}$	p
$^{72}_{32}\text{Ge}$	27.4	0.98	1.0	1.0	$^{72}_{32}\text{Ge}(n,\gamma)^{73}_{32}\text{Ge}$	
$^{73}_{32}\text{Ge}$	7.8	14.0	14.0	15.0	$^{73}_{32}\text{Ge}(n,\gamma)^{74}_{32}\text{Ge}$	
$^{74}_{32}\text{Ge}$	36.5	0.62	0.5	0.52	$^{74}_{32}\text{Ge}(n,\gamma)^{75}_{32}\text{Ge} \xrightarrow[82.8m]{\beta^-} {}^{75}_{33}\text{As}$	n
$^{76}_{32}\text{Ge}$	7.8	0.36	0.2	0.16	$^{76}_{32}\text{Ge}(n,\gamma)^{77}_{32}\text{Ge} \xrightarrow[11.3h]{\beta^-} {}^{77}_{33}\text{As} \xrightarrow[38.8h]{\beta^-} {}^{77}_{34}\text{Se}$	n

phosphorus donor is produced for every 10^{11} gallium majority acceptors during the NTD process. In other words: ultra-pure germanium crystals are virtually perfect starting material. For the present NTD study, we have chosen an ultra-pure germanium single crystal which we have grown at the crystal growth facility at the Lawrence Berkeley Laboratory.[15] Our ultra-pure germanium crystals are typically grown for nuclear radiation detectors i.e., large volume p-i-n junctions which are operated at liquid nitrogen temperature.[16] The specific crystal which was chosen for the study (LBL #516) has been grown in a hydrogen atmosphere (1 atm) from a melt contained in a pyrolytic carbon-coated quartz crucible using the Czochralski technique.[17] Six 2 mm thick slices of 36 mm diameter were cut, lapped and chemically etched. They were irradiated with thermal neutrons at the University of Missouri Research Reactor facility[18] with neutron doses ranging from 7.5×10^{16} to 1.88×10^{18} cm^{-2}. After more than ten half lives of $^{71}_{32}$Ge ($T_{\frac{1}{2}} = 11.2$d), the samples were annealed at $400°C$ for six hours in a pure argon atmosphere (1 atm) to remove the irradiation damage.

ELECTRICAL MEASUREMENTS

A few small square specimens (7×7 mm^2) were cut from each slice. The Van der Pauw geometry[19] was chosen for Hall effect and resistivity measurements. This geometry is very insensitive to the shape, the position and the size of the electrical contacts. In order to obtain degenerately doped p^{++} contacts on all four corners of both faces, we used the well established contact fabrication technology which was developed for Ge:Ga photoconductors.[20] These contacts are doped so highly that they are virtually metallic and can inject and extract holes down to the lowest measurement temperatures. The following is an abbreviated description of the contact fabrication process.

Both square faces were implanted with 25 and 50 keV boron ions at doses of 10^{14} and 2×10^{14}cm^{-2} respectively. Annealing at $250°C$ for one hour was followed by RF sputtering of 400 Å of titanium and 8000 Å of gold. The stress in the metal film was relieved by a second annealing step at $220°C$ for 0.5 hours. Areas of about $2 - 3$ mm^2 were covered with an etch resistant wax in all four corners on both faces. The gold layer was etched off with a KI + I$_2$ solution. The titanium layer was removed with a brief soaking in 1% HF in water. The implanted layer was removed by a 20 sec etching cycle in a HF:HNO$_3$ (1:3) mixture. After washing off the wax in trichloroethylene and rinsing in transistor-grade methanol, the samples were ready for measurement.

Resistivity (ρ) and Hall effect (R_H) measurements were performed in a liquid helium flow-type cryostat.[21] The results of both measurements yield the mobility μ:

$$\rho = (p\mu e)^{-1} \qquad (p = \text{free hole concentration, } e = \text{charge} \quad (15)$$
$$\text{of the electron)}$$

$$R_H = (pe)^{-1} \qquad\qquad\qquad\qquad\qquad\qquad\qquad (16)$$

$$\mu = R_H/\rho \qquad\qquad\qquad\qquad\qquad\qquad\qquad\qquad (17)$$

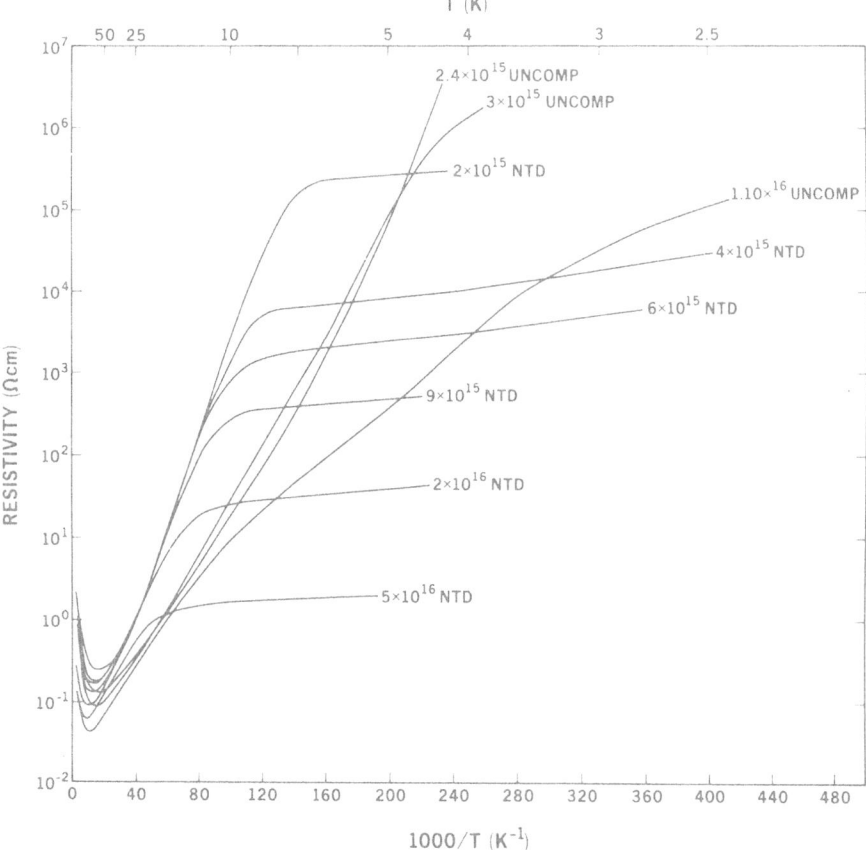

Fig. 4. Resistivity as a function of 1000/T for NTD and uncompensated germanium samples. Each curve is labeled by the
gallium concentration obtained by either NTD or melt
doping.

The mobility values are only useful down to the temperature where
hopping conduction sets in. Our mobility values agree very well
with published values for melt-doped material in the temperature
range above the hopping regime. This indicates that the concentration of residual radiation damage or other free-carrier scattering centers must be very small. Figure 4 shows the log (resistivity) versus 1000/T dependence for the six NTD germanium samples.
The number next to each curve corresponds to the acceptor (gallium)
concentration in each sample. For comparison we have also measured
gallium-doped germanium samples which have extremely small values
of K. These so-called uncompensated samples were cut from crystals
which were doped in the melt and were grown in the ultra-pure germanium crystal-growing equipment, and not NTD doped. The compensating donor concentration in these crystals is estimated to be
less than 10^{11} to 10^{12} cm^{-3}. The resulting K is of the order of
10^{-4} to 10^{-5}.

The resistivity-temperature dependence of these NTD samples is
characterized by three regimes. At high temperatures (room temperature down to about 50 K), the resistivity decreases because the
carrier mobility increases. Below about 50 K carrier freeze out
begins and reduces the free hole concentration rapidly. The slope
of the freeze out in highly-compensated material is proportional to
the acceptor binding energy $E_A - E_V \simeq 11$ meV. At still lower temperatures, the appearance of hopping conduction causes the resistivity to increase only very slowly. All six NTD germanium samples
show these three resistivity regimes very clearly. The low-compen-
sation samples show different log (ρ) versus 1/T dependences. In
these samples the concentration of ionized acceptors is low, and
hopping from neutral to ionized centers is improbable. A third
conduction mechanism has been proposed for such material.[8] It is
based on the idea that carriers can "hop" from a neutral to a
neighboring neutral acceptor thereby forming a positively charged
acceptor. The carrier binding energy for such positive acceptors
is very small, leading to the characteristic conduction range
between 10 and 3.5 K in the sample 1.1 x 10^{16} uncomp.

The samples which show a resistivity dependence which is of
interest to bolometer applications are the three most highly doped
NTD crystals. They were measured to temperatures as low as
T = 0.3 K. Figure 5 shows the same dependence as Fig. 4 but the
1/T axis is about tenfold compressed. The resistivity can be
expressed empirically with the following equation:

$$\rho = \rho_0 \, \exp(\Delta/T) \tag{17}$$

The following values for ρ_0 and Δ have been fitted to the experimental curves of Fig. 5:

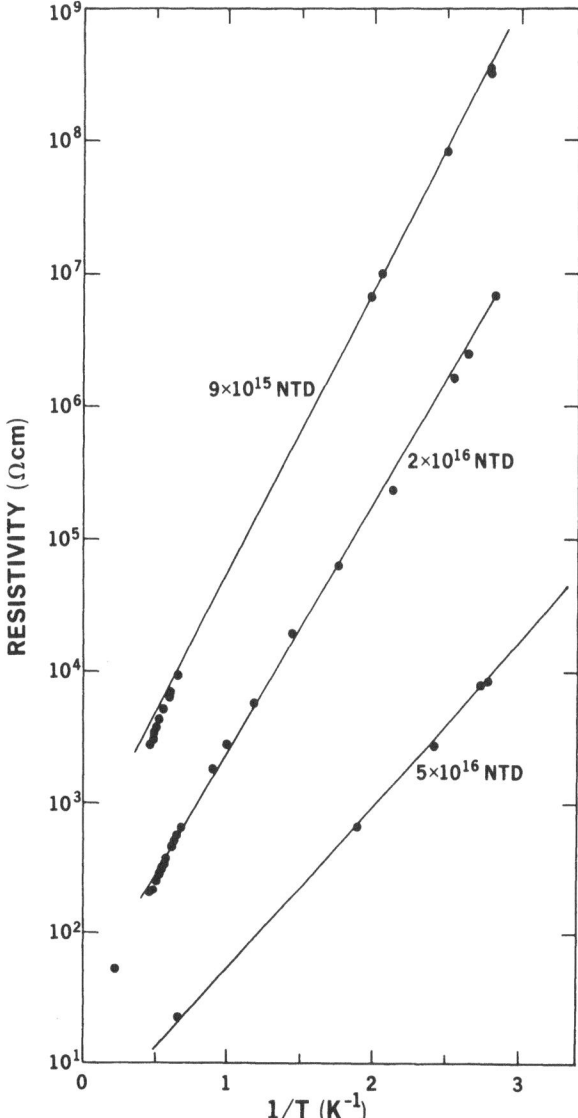

Fig. 5. Resistivity as a function of 1/T for the most heavily-
         doped NTD germanium samples.

NTD Sample	$\rho_0(\Omega cm)$	$\Delta(K)$
9×10^{15}	430.0	4.9
2×10^{16}	34.0	4.4
5×10^{16}	3.3	2.8

Depending on the desired bolometer operating temperature, the bolometer dimensions and the optimum resistance value, one can virtually produce the ideal doping concentration for any case. The sample
5×10^{16} has been used for the fabrication of bolometers. These
devices have been field tested and have been found to work as well
and as reliably as any conventional bolometer. The noise for these
field tests was limited by effects not related to the bolometer.
Critical tests on the noise as a function of the bolometer measuring current have not been performed so far but are planned for the
near future.

CONCLUSIONS AND SUMMARY

    Neutron transmutation doping seems to be the ideal doping tech-
nology for germanium which is used for low temperature bolometers.
The excellent homogeneity of the doping concentration and compensation produced by the NTD process is, for this particular application, even more important than in the typical application of NTD
silicon for high-voltage, high-power devices. In the low temperature range, conduction mechanisms begin to dominate which rely on
quantum mechanical tunneling (i.e., "hopping") of charge carriers
from one acceptor site to an empty neighbor site. The tunneling
probability is exponentially dependent on the interimpurity distance which explains the extreme sensitivity of hopping conduction
to the homogeneity of the dopant distribution.

    The NTD process in high-purity germanium leads to a fixed
compensation which in turn results in a certain slope of the log
(ρ) versus 1/T dependence (see Fig. 5) for a given neutron exposure. This slope represents the temperature coefficient α of the
material. Using Eq. (7) and (17) we find:

$$\alpha = \frac{1}{R} \ (dR/dT) = \frac{1}{\rho} \ (d\rho/dT) = -\Delta/T^2 \tag{18}$$

Because the responsivity is directly proportional to α (Eq. 9), we
see that a large value of Δ and a small value of T are desirable.
The Δ values are a factor of 1.5 to 2.0 smaller than the values of
the best bolometer material for T = 4.2 K applications. The noise
equivalent power--the most important figure of merit--depends only

on the square root of Δ which renders the difference between NTD
material and the conventional material negligible. At lower temperatures, the difference vanishes and at the lowest temperatures
no conventional material has been found which would be useful.
These facts together with the possibility of producing reliably
large quantities of predictably, homogenously-doped germanium
using the NTD process seem to have created a good solution to the
fabrication of low temperature bolometer material.

The two tasks which remain to be performed for NTD germanium
are on the one hand, the accurate determination of the thermal
neutron cross sections and on the other hand, the fine tuning of
the NTD process for the production of material which is ideally
suited for several typical bolometer operating temperatures (e.g.
3, 1.5, 0.3 and 0.1 K). A careful series of noise tests have to
be performed at these temperatures as a function of excitation
current in the bolometer to determine if there are any excess
noise sources present in this material.

ACKNOWLEDGEMENTS

We are indebted to J. Meese for his help in performing the neutron irradiations and to F.S. Goulding for his continued interest.

REFERENCES

1. E. H. Putley "Topics in Applied Physics, Volume 19: Optical
    and Infrared Detectors," R. J. Keyes, ed., Springer-Verlag,
    New York (1977).
2. F. J. Low, J. Opt. Soc. Amer., 51:1300 (1961).
3. R. G. Rhodes, "Imperfections and Active Centres in Semiconduc-
    tors," The MacMillan Co., New York (1964).
4. P. M. Downey, Ph.D. thesis, Dept. of Physics, Massachusetts
    Institute of Technology (1980).
5. H. Fritzsche and M. Cuevas, Phys. Rev., 119:1238 (1960).
6. N. S. Nishioka, P. L. Richards, and D. P. Woody, Appl. Opt.,
    17:1562 (1978).
7. N. F. Mott, "Metal-Insulator Transitions," Barnes and Nobles
    Books, New York (1974).
8. H. Fritzsche, "The Metal Non-Metal Transition in Disordered
    Systems," L. R. Friedman and D. P. Tunstall, eds., Scot-
    tish Universities Summer School in Physics, St. Andrews,
    Scotland (1978).
9. "Neutron Transmutation Doping in Semiconductors," J. Meese,
    ed., Plenum Press, New York (1979).
10. "Neutron-Transmutation-Doped Silicon," J. Guldberg, ed.,
    Plenum Press, New York (1981).
11. H. C. Schweinler, J. Appl. Phys., 30:1125 (1959).

12. "Table of Isotopes," Sixth Edition, C. M. Lederer, J. M. Hol-
     lander, and I. Perlman, eds., John Wiley and Sons, Inc.,
     New York (1967).
13. "Table of Isotopes," Seventh Edition, C. M. Lederer and V. S.
     Shirley, eds., John Wiley and Sons, Inc., New York (1978).
14. E. E. Haller, W. L. Hansen, and F.S. Goulding, Adv. Phys., 30,
     No. 1, 93 (1981).
15. Dept. of Instrument Science and Engineering, Lawrence Berkeley
     Laboratory, University of California, Berkeley, California
     94720.
16. E. E. Haller and F. S. Goulding, Nuclear Radiation Detectors,
     in: "Handbook on Semiconductors," Vol. 4, C. Hilsum, ed.,
     North Holland Publishing Co. (1981).
17. W. L. Hansen, Nucl. Instr. and Methods, 94:377 (1971).
18. J. Meese, University of Missouri Research Reactor Facility.
19. L. J. van der Pauw, Philips Res. Repts., 13:1 (1958).
20. NASA Progress Reports, Contract W-14,606. Copies available
     from E. E. Haller, Lawrence Berkeley Laboratory.
21. Helium cryostat Model 310 Cryotran Refrigerator, Lakeshore
     Electronics Inc., Westerville, Ohio 43081.

RELIABLE IDENTIFICATION OF RESIDUAL DONORS IN HIGH PURITY

EPITAXIAL GALLIUM ARSENIDE BY TRANSMUTATION DOPING[*]

M. N. Afsar

Massachusetts Institute of Technology
Francis Bitter National Magnet Laboratory[†]
Cambridge, Massachusetts 02139

ABSTRACT

Neutron transmutation doping of n-type Gallium Arsenide is not new but it has recently proved to be essential for the identification of residual donor contaminants in high-purity n-GaAs. The hydrogen-like donors lead, silicon, selenium, tin, tellurium, sulfur, germanium and carbon can be observed seperately at low temperature because each bound donor electron has a different ionization energy. Transmutation doping with slow neutrons provided the first unquestionable identification of selenium and germanium which now permits unknown specimens of n-GaAs to be examined for the presence of these elements.

INTRODUCTION

Attempts to introduce specific donor impurities chemically into GaAs always seem to result in the concomitant introduction of other unknown donors,an increase in acceptor concentration and sometimes the introduction of the desired donor. Chemical back-doping has been so unreliable that it has misled spectroscopists[1-6] for a decade in their attempts to identify the specific donor Pb, Si, Se, Sn, S, Te, Ge and carbon. A very reliable method for experimental observation of each of these donors seperately has existed for more than a decade[4] namely,the far infrared photoconductivity technique.

---

[*] Work supported by the U.S. Air Force Office of Scientific Research under Grant No.AFOSR-78-3708A.
[†] Supported by the National Science Foundation

At low temperature the donor electron is excited from the hydrogen-
like ground state by far infrared photons. Since the ionization
energy of the bound electron is different for each of these different donor nuclei, it is not difficult to see a separate excitation
for each donor. It has not been possible for many years, however,
to specify which of the observed excitations is to be associated
with each donor species. Neutron transmutation doping with selenium
and germanium gave us one of the first unquestionable identification
of these two donors.[7-9] Since then, at least one other method has
been used [10-12] for the unquestionable identification of some of
the other donors, for example from the study of molecular beam epitaxy (mbe) grown Gallium Arsenide.

The (n, γ) reaction with the stable Gallium and Arsenide nuclides leads to the formation of the following impurity centers[10]

$$^{69}Ga \; (n, \; \gamma) \, ^{70}Ga \xrightarrow{\;\;\beta\;\;} \, ^{70}Ge,$$

$$^{71}Ga \; (n, \; \gamma) \, ^{72}Ga \xrightarrow{\;\;\beta\;\;} \, ^{72}Ge,$$

$$^{75}As \; (n, \; \gamma) \, ^{76}As \xrightarrow{\;\;\beta\;\;} \, ^{76}Se.$$

Since the (n, α), (n, p) and $(n, 2n)$ reactions, as well as the
fast neutron reactions, can be neglected [13], the effect of transmutation doping of GaAs is to introduce germanium and selenium impurities. When we examined the photoconductivity spectrum of the
neutron irradiated epitaxial GaAs and compared it to that of the
virgin specimen from which it was cut, we found two extra excitation
peaks to which we were compelled, by reason, to assign the identity
of selenium and germanium. These identifications can now serve as
the universal basis for detecting residual, unintentional contamination of epitaxial GaAs by selenium and germanium whenever n-type
material needs to be characterized. The method by which the spectra
of any new specimen may be compared to the present, original observations for the purpose of detecting the contamination by selenium
or germanium will now be described.

PHOTOCONDUCTIVITY IDENTIFICATION

The far-infrared photoconductivity method is virtually the
only way to observe separately the various donor species in ultrahigh purity epitaxial GaAs. Such a low concentration ($\sim 10^{13}$ to
10^{14} cm^{-3}) of residual, unintentional contaminants can only be resolved from each other in the presence of a high intensity magnetic
field.

In the case of a hydrogen-like donor in gallium arsenide, the
single bound electron in the 1s ground state at low temperatures

can be excited to a 2p state by the absorption of a photon. Such
a high state will be near enough to the conduction band to permit
the bound electron then to be thermally ionized, thus allowing the
original photon absorption to be detected by the far-infrared photoconductivity method. The 1s → 2p transition has a different transition energy for different donor nuclei, because there is a probability for the electron to be near to the nucleus when it is in the
1s ground state. An applied magnetic field removes the threefold
degeneracy of the excited 2p state and has a measurable effect on
the 1s state, mainly that of compressing it so that the electron
is influenced more strongly by the nature of the nucleus to which
it is bound. This means that a high intensity magnetic field provides three advantages. First, the threefold degeneracy of the 2p
state is removed and the relatively narrow transition 1s → 2p (m=-1)
can be observed. Second, higher intensity magnetic fields cause the
narrowing of each line and the third, the spreading apart of lines
as shown in Fig. 1. In short, the lines are resolved at high fields.
The observation of any one of the 1s → 2p (m = -1, 0, +1) transitions
is adequate for the reliable identification of donor impurities. It
is preferable to observe 1s → 2p (m = -1) transition since lines
are narrow for this transition compared to other transitions.

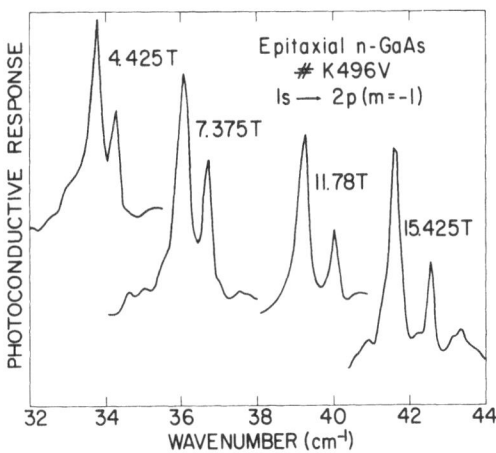

Figure 1. Photoconductivity spectra of 1s → 2p (m = -1) transition
at magnetic field intensities of 4.425 T, 7.375 T, 11.78 T and
15.425 T obtained with a vapor-phase epitaxy-grown specimen. The
lines get narrower and spread apart with increase in magnetic field
intensity. In a recent study it is now revealed that the line on
the left is attributed to silicon donor and the line on the right
hand side is attributed to sulfur [10,11,15]

INTERFEROMETRIC SPECTROMETER

A modern high-resolution modular cube-type Fourier transform spectrometer has been used for the photoconductivity measurements.[8,9] It uses phase modulation by oscillating the mirror in one of the active arms of the interferometer, rather than a conventional chopper-operated amplitude modulation for improved (~4 times more) signal to noise ratio. A typical resolution of the spectrometer is 0.05 cm^{-1}, but it can easily be increased to 0.025 cm^{-1}. The broad band source is a medium-pressure quartz-encapsulated mercury-vapor lamp. In the photo conductive mode, the specimen acts as its own detector. The output beam of the interferometer is focused on to a light pipe which guides the beam to the specimen resting in the tail of a cryostat which in turn extends into the bore of a water-cooled copper Bitter solenoid magnet. In our measurement the specimen is mounted vertically (Voigt configuration, i.e., electric bias field parallel to the magnetic field). In this configuration, other Rydberg and Zeeman transitions could also be observed when necessary for fundamental studies. The specimen temperature is increased from 4 K to up to about 8 K in steps to reduce the magnetoresistance at high fields for better signal to noise ratio.

GENERATION OF THE SIGNATURE CURVE

Figure 1 shows the photoconductive response for the 1s → 2p (m = -1) transition for the donor electron of silicon (left peak) and sulfur (right peak). These two donors always appear in the highest purity specimens grown by the method of vapor phase epitaxy. Note that the spectral lines become narrower and become resolved as the magnetic field intensity is increased from 4 Teslas to 15 Teslas. For many years, these two lines were attributed to a "gallium vacancy"[3] and a silicon donor,[4] respectively. Because the quartz vapor-phase vessel is expected to contribute a silicon contaminant and the arsenic starting material may contribute a sulfur contaminant, it is not surprising that Ozeki [11,12] demonstrated the proper identification recently. A plot of the peak positions of these transitions shown in Fig. 1 at each value of magnetic field develops a "signature curve" for each of the donor contaminants. Indeed, the signature curves of Fig. 2 show that we are dealing with the electrons of two hydrogen-like donors and not the 1s → 2p (m = -1) transition of a stoichiometric defect.

It is from this virgin specimen, #K496V that material was taken and bombarded with slow neutrons and redisignated as specimen #K496B. The signature curves of Figure 2 are needed to establish the true donor nature of the two additional spectral lines that appear in #K496B. It is also necessary to establish their relative position in the energy spectrum with respect to silicon and sulfur.

SPECIMEN PREPARATION

A few samples with surface dimensions 5 mm x 6 mm were cut from wafer #K496V of epitaxial GaAs grown by vapor phase epitaxy on

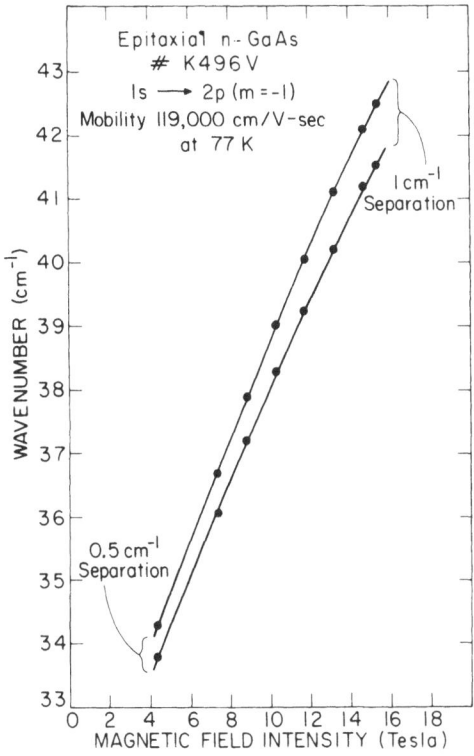

Figure 2. Plots of peak positions (the signature curve) as a function of applied magnetic field (8 different magnetic fields) for two donor contaminants of Fig. 1. The signature curves separate as the magnetic field is increased.

a semi insulating Cr- doped substrate. The thickness of the epitaxial layer was 51μm and the substrate thickness was 340μm. The mobility and carrier concentration at 77 K were 119,000 cm^2/v-sec and 3.7 x 10^{14}/cm^3, respectively. The MIT Nuclear Reactor provided the neutrons for the bombardment of two of these specimens. The bombardment time was 79 minutes at a flux density of 10^{12} neutrons/cm^2-sec with a nominal ratio of "slow" to "fast" neutrons of about 100. After the bombardment, the samples were stored for over three weeks, to decay to a negligible level of radioactivity. One of the sample was annealed in nitrogen and the other in hydrogen, each for 15 minutes at 600 C. The liquid nitrogen temperature mobility and donor concentration changed to 89,500 cm^2/V-sec, 8.3 x 10^{14}/cm^3 for the specimen annealed in nitrogen and to 91,700 cm^2/V-sec, 8.7 x 10^{14}/cm^3 for the specimen annealed in hydrogen.

In this paper we are reporting results obtained with only the bombarded specimen annealed in nitrogen, the K496B together with unbombarded virgin specimen K496V. We have not made any detailed study of the specimen annealed in hydrogen yet. Electrical connections were provided by indium coated platinum contact leads at four corners of each specimen.

TRANSMUTED SELENIUM AND GERMANIUM

The upper part of Fig. 3 shows the silicon and sulfur lines

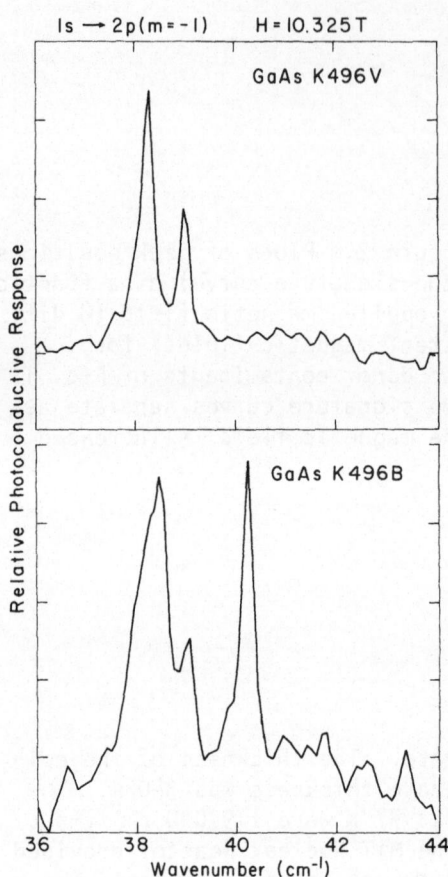

Figure 3. Photoconductivity spectra for 1s → 2p (m = -1) transition of virgin and bombarded n-GaAs at magnetic field intensity of 10.325 T. The spectrum of the bombarded specimen shows an additional donor (Ge) on the right and the presence of another donor (Se) on the left in the unresolved broad line.

of the virgin specimen. The spectrum of the bombarded specimen shown in the lower part of Fig. 3 exhibits two extra features. At the far lower right is the sharp line introduced by the transmutation of gallium to germanium. The lower left line broadened, indicating the presence of unresolved selenium line.

Calculations of central cell corrections indicate that the germanium transition should have the highest energy of the group except for carbon. (Since carbon is never seen as a donor, only as an acceptor, spectral lines appearing in this position in the past have been attributed to carbon simply because no line was ever seen at higher energy).

The selenium spectral line appears unresolved between silicon and sulfur in the lower part of Fig. 3. The higher magnetic field helps with the observations as shown in Fig. 4 where the sulfur

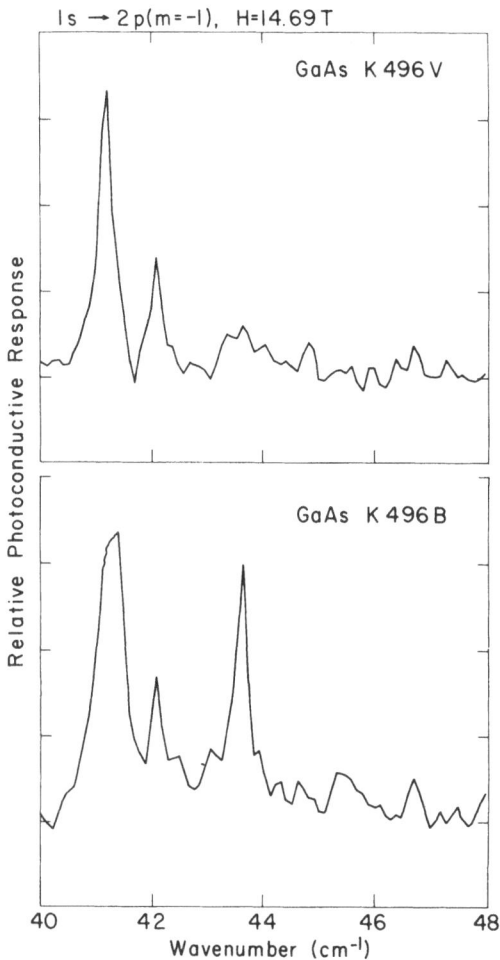

Figure 4. Photoconductivity spectra for 1s → 2p (m = -1) transition obtained with virgin and bombarded specimens at magnetic field intensity of 14.69 T. Spectrum of the bombarded specimen exhibits one additional donor (Ge) on the right. The other transmutation doped donor (Se) broadened the usual silicon line on the left.

line has been resolved leaving a composite silicon and selenium transition at the far left.

Figure 5 shows the two spectra of Fig. 4 superimposed at a field of 14.69 Teslas. The line on the far left has broadened toward higher energy indicating the presence of the selenium line as claimed above when the two spectra of Fig. 5 are subtracted, the two lines shown in Fig. 6 are obtained. The final test was the plot of the peaks of these two lines observed at different values of magnetic field intensity generates the signature curves shown in Fig.7.

Two more tasks remain to be attempted, namely, (1) the re-examination of spectra from previous specimens to compare the original attributions to these unambiguous identifications, and (2) the identification of the remaining donors such as tin.

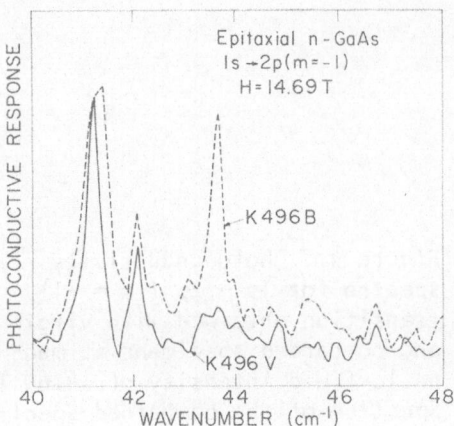

Figure 5. Spectra of Fig. 4 at 14.69 T superimposed to show the broadening of the line on left and the presence of transmuted selenium line. The sulfur lines at 42 cm^{-1} are in exact coincidence indicating that they are resolved. When the spectrum obtained with the virgin specimen is subtracted from the bombarded (dashed), the selenium line appears as shown in Fig. 6.

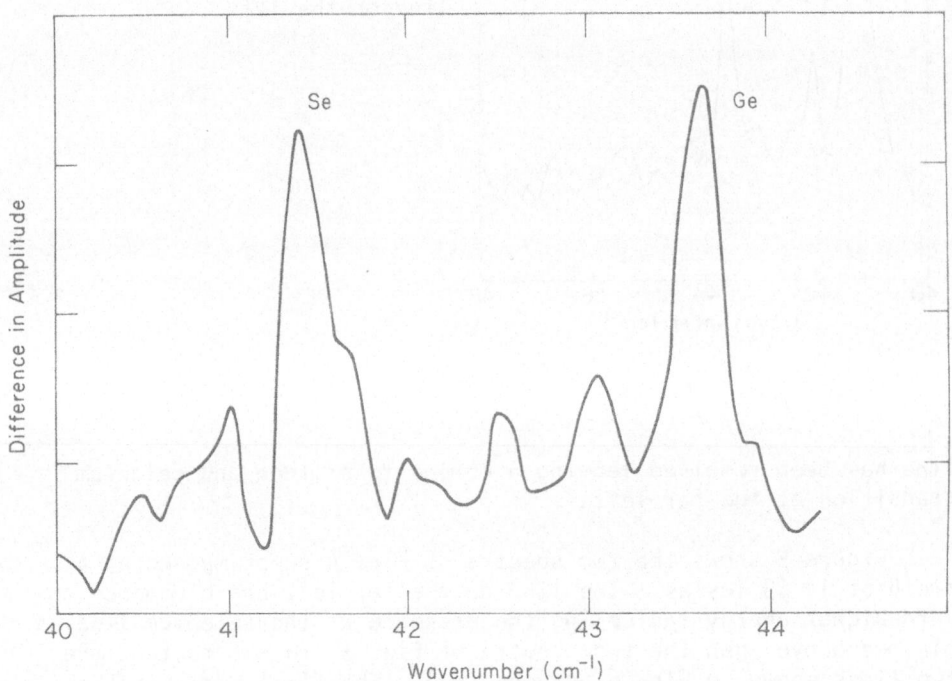

Figure 6. The solid line of Fig. 5 was subtracted from the dashed line to resolve further the selenium and germanium lines.

Figure 8 shows the spectra for a Lincoln Laboratory specimen # B4-21A which appears to contain four unintentional residual donors. When the spectra were reproduced at different values of magnetic field, the peak of each line was plotted as an (x) in Fig. 9.

It is clear that the "unknown" test specimen contains silicon, selenium, sulfur and germanium, reading from left to right in Fig. 8. These contaminants had previously been attributed to a gallium vacancy, tin, silicon and carbon respectively.[3]

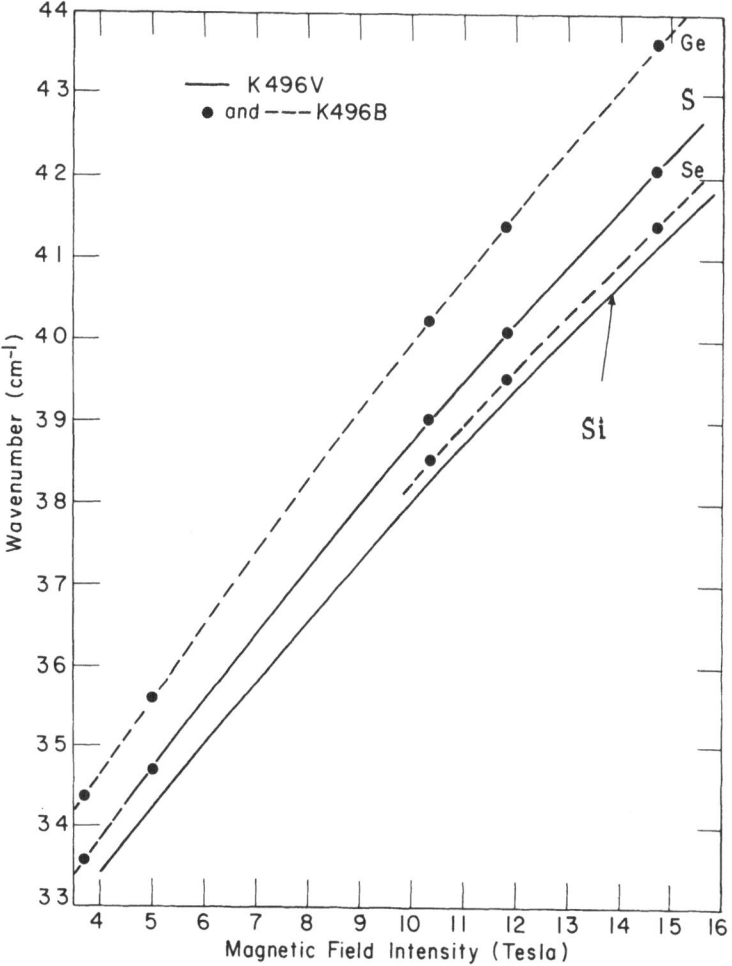

Figure 7. Signature curves for germanium and selenium donor levels generated by plotting peak positions of subtracted spectra at different magnetic fields. Full lines represent signature curves of silicon and sulfur generated by the unbombarded virgin specimen shown earlier in Fig. 2.

A METHOD FOR IDENTIFYING THE REMAINING DONOR SPECIES

When very high quality GaAs is prepared by the method of molecular beam epitaxy (mbe), the layer is p-type. It can be deliberately prepared as n-type material by admitting a donor during growth. The donor tin was chosen to demonstrate this.[14]

Because the specimen is of extremely high quality, only the intentionally added tin donor is observable, the others being too weak to detect. A signature curve for tin was developed in this fashion. The original identification of tin by Fetterman, et al [16] with an NH_3 laser was found to lie on our signature curve correct to four significant figures both in magnetic field intensity and in the energy of the 1s → 2p (m = -1) transition (Fig. 10).

RELATIVE DONOR CONCENTRATION

There is no reason to expect that the phtoconductivity method can provide a means for the precise measurement of donor concentration. The high precision of this method lies in the determination of the energy of quantum transitions.

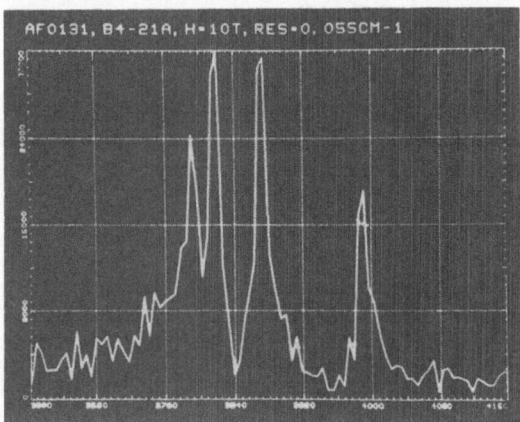

Figure 8. Our photoconductivity spectrum for 1s → 2p (m = -1) transition obtained with the Lincoln Laboratory unintentionally doped vapor phase epitaxy grown specimen B4-21A (μ_{77}=210,000 cm^2/V-sec, n_D=0.5x10^{14} cm^{-3} and n_A = 0.2x10^{14} cm^{-3} at 10 T.

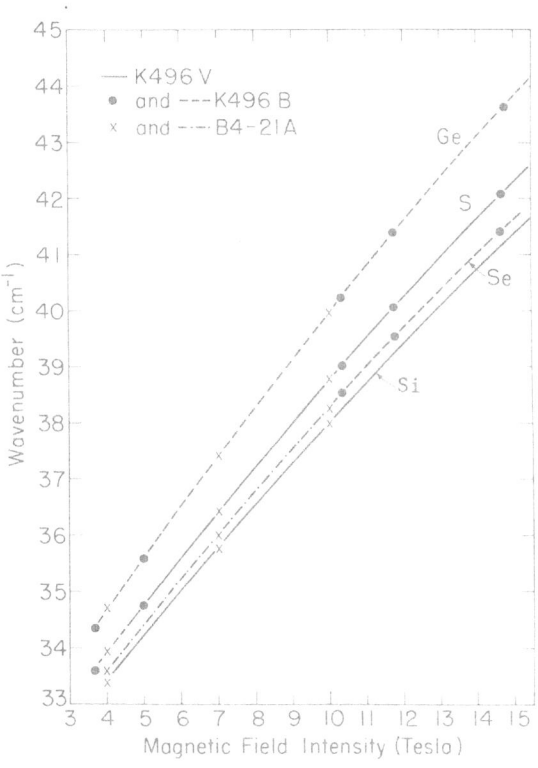

Figure 9. Correct characterization of donor impurities in the
vapor phase epitaxy grown specimen B4-21A. By comparing signature
curves generated by spectra of B4-21A with signature curves generated by spectra of neutron-doped specimen K496B, additional donors
in the unintentionally doped vapor phase epitaxy grown specimen
B4-21A are identified as germanium and selenium. The x's obtained
from peak positions at magnetic field intensity of 4, 7, and 10
Tesla with specimen B4-21A fall on four signature curves generated
by specimens K496V and K496B.

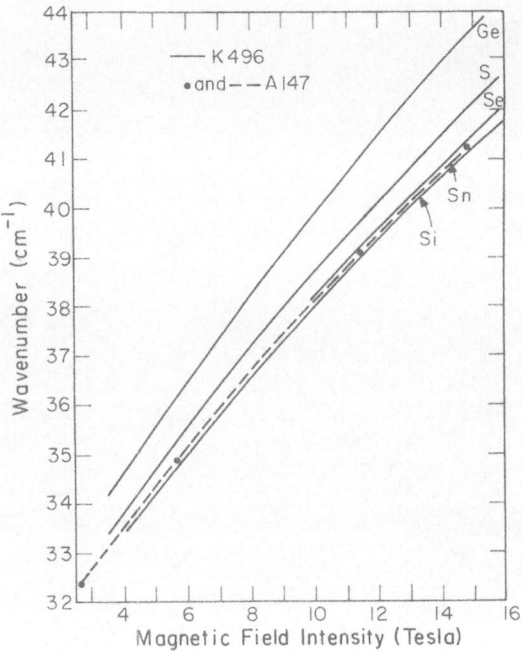

Figure 10.   Signature curve for tin donor obtained with molecular beam epitaxy grown specimen #A147 shown together with signature curves for germanium, silicon, selenium and sulfur donors obtained with the neutron irradiated vapor phase epitaxy grown specimen #K496.

The line shape of these transitions is discussed in another article.[17] Close examination of low-field transitions shows an abrupt edge on the high energy side of the line and a long tail on the low energy side of the line. This tail is suppressed in ultrahigh magnetic fields. Nevertheless, it is not possible to compare line intensities to better than a factor of five when the line shape is not understood and is not statistically tractable. The area under the curve cannot be calculated or even measured in the presence of a long tail.

Therefore, the height of different peaks can be expected to indicate differences of an order of magnitude in concentration, if it exists, but the relative height of peaks cannot be expected to correlate with capture cross sections that differ by 50%.

CONCLUSIONS

Because the residual, unintentional contaminants selenium and germanium in high purity n-GaAs had been identified as tin and carbon, it was necessary to employ transmutation doping to generate signature curves for these two donors, germanium and selenium as shown in Fig. 7. We have demonstrated that it is now quite simple for anyone to examine a specimen containing several unknown species of donors and identify germanium and selenium as contaminants if they are present as shown in Fig. 9.

Future possibilities for transmutation doping in this research program are abundant. For example, high quality epitaxial GaAs prepared by modern molecular beam epitaxy (mbe) always comes out to be p-type. The best way to examine this material for residual, unintentional donor contaminants is by transmutation doping with donors to turn the specimen to n-type. We are ready to employ this method of characterization of mbe material because we are generating signature curves for all other donors by intentionally doping mbe specimens with a single donor during growth.

We have just begun to examine the common ternary compounds such as $In_{0.53}Ga_{0.47}As$ and have preliminary signature curves for two donors. It is clear that the present transmutation doping procedure may be applied directly to this class of materials to generate signature curves for tin, germanium and arsenic. In the case of Indium Phosphide and the ternary phosphides, the signature curve for sulfur can also be generated.

ACKNOWLEDGMENTS

The author is grateful to Dr. Gerald L. Witt for his encouragement in this work. He also wishes to thank his colleagues, Dr. Kenneth J. Button, Dr. Steven H. Groves and Dr. David M. Larsen

for continuing advice, assistance and specimens and Dr. A. Y. Cho
for mbe specimens of highest quality.

REFERENCES

1.   G. E. Stillman, C. M. Wolfe and J. O. Dimmock, Solid State
      Commun., 7, 921 (1969).

2.   G. E. Stillman, D. M. Larsen and C. M. Wolfe, Phys. Rev. Lett.
      27, 988 (1971).

3.   G. E. Stillman, C. M. Wolfe and D. M. Korn, Proc. 13 th Int.
      Conf. Phys. Semicond., Rome 1976, pp 623-626.

4.   C. M. Wolfe, D. M. Korn and G. E. Stillman, Appl. Phys. Lett.,
      24, 78 (1974).

5.   R. A. Stradling, L. Eaves, R. A. Hoult, N. Miura and C. C.
      Bradley, Gallium Arsenide and Related Compounds 1972, (Inst.
      Phys. Conf. Ser. 17) pp 65-74

6.   R. A. Cooke, R. A. Hoult, R. F. Kirkman and R. A. Stradling,
      J. Phys. D, Appl. Phys. 11, 945 (1978).

7.   J. H. M. Stoelinga, D. M. Larsen, W. Walukiewicz, R. L.
      Aggarwal and C. O. Bozler, J. Phys. Chem. Solids 39, 873
      (1978)

8.   M. N. Afsar and K. J. Button, 4th. Int. Conf. on IR & MM Waves,
      Conf. Digest., IEEE Cat No. 79CH 1384-7 MTT, pp 10-11 (1979).

9.   M. N. Afsar and K. J. Button and G. L. McCoy, Inst. Phys. Conf.
      Ser. No.56, Chapter 8, p 547 (1981).

10.   M. N. Afsar, Microelectronics Journal, Special issue on Gallium
      Arsenide 13, 15 (1982).

11.   M. Ozeki, K. Kitihara and K. Nakai, Fijitsu Sci. and Tech.
      Jour., p 77 (June 1977).

12.   M. Ozeki, K. Kitihara, K. Nakai, A. Shibatomi, K. Dazai, S.
      Okawa and O. Ryuzan, Japan J. Appl. Phys. 16, 1617 (1977).

13.   Sh. M. Mirianashvili and D. I. Nanobashvili, Sov. Phys. Semi-
      cond. 4, 1612 (1971).

14.   M. N. Afsar, K. J. Button, A. Y. Cho and H. Morkoc, Int. J.
      IR & MM Waves, 2, 1113 (1981).

15. C. W. Litton, Private communication

16. H. R. Fetterman, D. M. Larsen, G. E. Stillman, P. E. Tannenwald
     and J. Waldman, Phys. Rev. Lett. $\underline{26}$, 975 (1971).

17. M. N. Afsar, K. J. Button and G. L. McCoy, Int. J. IR & MM
     Waves, $\underline{1}$, 513 (1980).

SPALLATION NEUTRON DAMAGE IN GROUP IV, III-V AND II-VI

SEMICONDUCTORS AT 5 K[*]

R. C. Birtcher, J. M. Meese[+] and T. L. Scott

Materials Science Division
Argonne National Laboratory
9700 South Cass Avenue
Argonne, Illinois 60439

ABSTRACT

Spallation neutrons from the Argonne Intense Pulsed Neutron Source have been used to investigate low temperature (5 K) neutron damage to a representative group of IV, III-V and II-VI semiconductors. Carrier removal rates and isochronal annealing are presented for all the semiconductors investigated while Deep-Level Transient Spectroscopic experiments on n-type silicon are discussed.

INTRODUCTION

Although cryogenic neutron irradiation facilities have been available in limited numbers at several steady-state reactors since the pioneering development of these facilities by Coltman, Blewitt and Noggle (1957)[1], virtually no data exists on degraded fission spectrum neutron damage to semiconductors near liquid-helium temperatures. Such experiments, which are highly desirable from a fundamental point of view, are nearly impossible to perform in steady-state reactors since the accompanying gamma flux produces an extraneous photoconductivity which interferes with the desired DC conductivity measurements used to obtain carrier removal rate and annealing data. Cleland and Crawford have

[*]Work supported by the U. S. Department of Energy.
[+]Permanent address: Research Reactor Facility, University of Missouri, Columbia, MO 65211.

reported neutron damage to degenerate (n_o ~10^{18} cm^{-3}) Germanium
irradiated in a steady-state reactor at 16 K. They observed that
the DC conductivity decreased by about a factor of 5 at the end of
the irradiation when the reactor was turned off due to the
associated reduction of the gamma flux.[2] They also concluded that
gamma flux after shutoff was sufficient to produce an appreciable
number of minority-carrier trapping centers.[2] Stein overcame
these difficulties by using neutrons from the Sandia pulsed
reactor (presumably measuring between pulses) to obtain carrier
removal rate and annealing data for non-degenerate n- and p-type
silicon and epitaxial n-type GaAs irradiated at 76 K.[3-5]

In the present experiments, these problems were solved by
using spallation neutrons from the Intense Pulsed Neutron Source
(IPNS) at Argonne National Laboratory. This accelerator-based
source provides bursts of neutrons and gammas so short compared to
the pulse duty cycle that continuous DC conductivity measurements
can be performed without any measurable interference. All irradiations were performed on degenerate n- and p-type silicon and on
degenerate n-type GaAs, CdS and CdTe. Deep Level Transient
Spectroscopic (DLTS) experiments on non-degenerate n-type silicon
irradiated at 5 K and isochronally annealed to above room
temperature will also be presented.

EXPERIMENTAL DETAILS

The details of the Argonne IPNS irradiation facility have
been described previously.[6] Spallation neutrons from this
accelerator-based source are produced by the interaction of
400-MeV protons from a rapid cycling synchrotron (RCS) with a ^{238}U
target. The RCS produces 100 ns wide proton pulses with a
repetition rate of up to 30 Hz. The proton beam can be directed
toward either of two targets, the radiation effects facility (REF)
target or the neutron scattering facility (NSF) target as shown in
Fig. 1. These two targets, located approximately 2.5 meters apart
are shielded from each other primarily by 2.5 meters of iron. On
either side of the REF target are two vertical irradiation
thimbles equipped with liquid-He cryostats fed from an elaborate
system of Collins-machines and compressors. A multiple-foil
activation analysis technique has been used to determine the
neutron energy spectrum[7] at the irradiation position and the
results are shown in Fig. 2. This figure shows a plot of the
neutron fluence per unit lethergy ($E \, d\phi_n/dE$) per proton incident
on the uranium target as a function of neutron energy. The
fluence of neutrons with E>0.1 MeV is ~1.5 x 10^{-2} n/cm^2/p while
the neutron fluence above the silicon displacement threshold
(E>0.188 keV) is ~2.0x10^{-2}n/cm^2/p. This number will be used to
convert from carriers removed per proton incident on the target
to carriers removed per n/cm^2 incident on the sample. The ratio

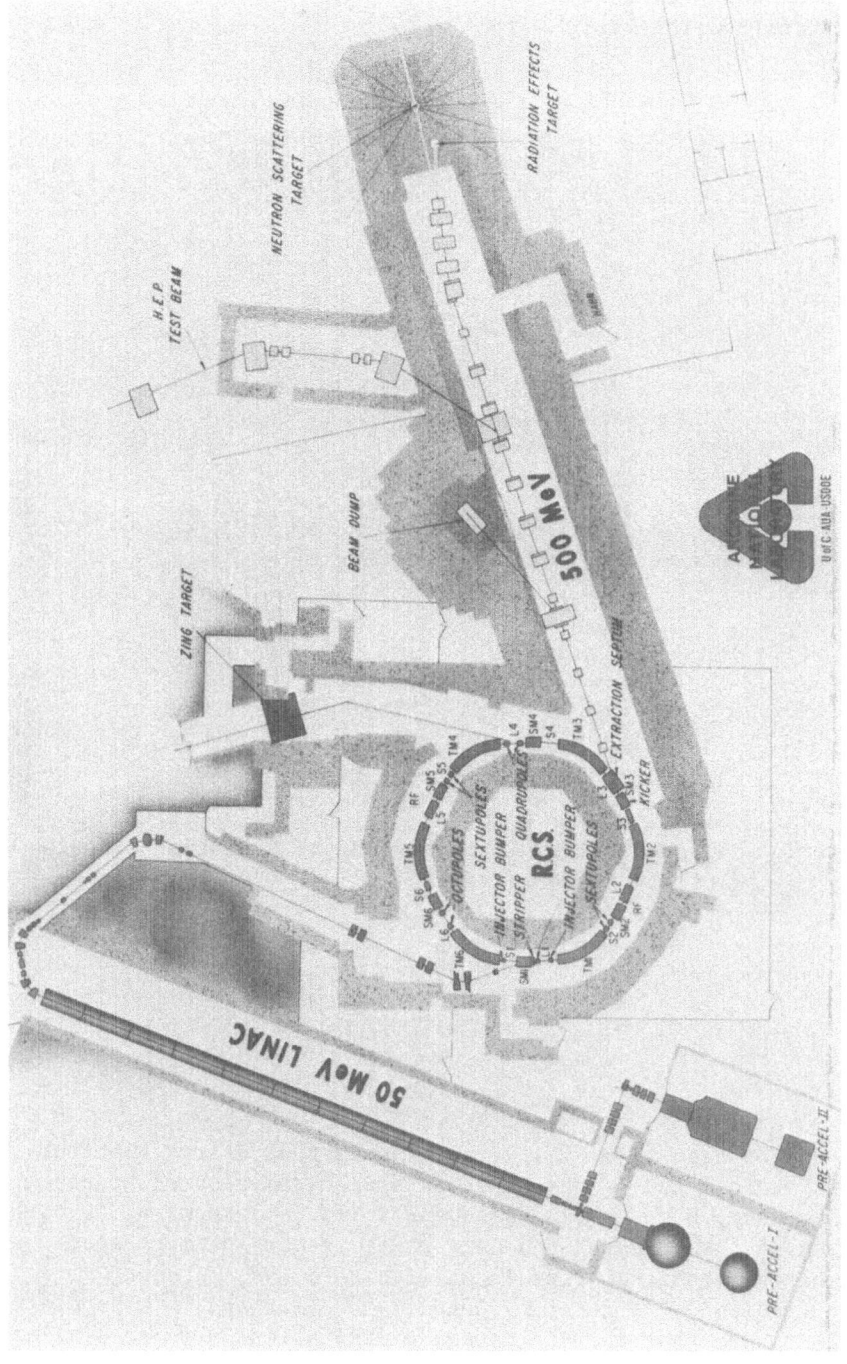

Figure 1. Schematic diagram of the Argonne Intense Pulsed Neutron Source (IPNS) showing the neutron scattering facility (NSF) and the radiation effects facility (REF).

of "fast" (E>0.1 MeV) to thermal neutrons is about 70. The
secondary proton flux is estimated to be about 0.4% of the flux of
neutrons with E>0.1 MeV.

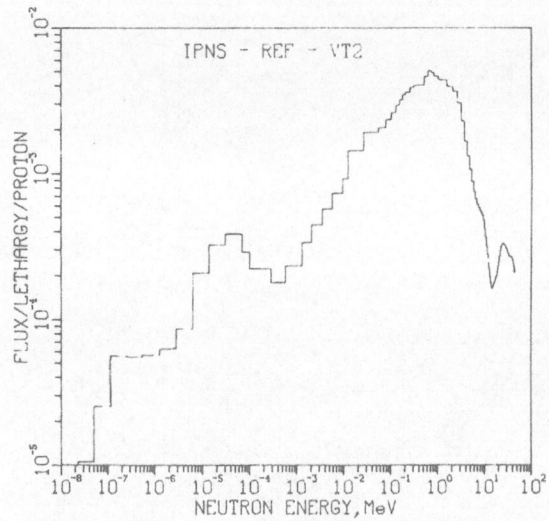

Fig. 2.   Neutron spectrum produced at the radiation effects
          facility target.

It can be seen from Fig. 2 that the neutron energy spectrum
at the REF target can be described as a degraded fission spectrum
with a low flux, high energy component. Figure 3 shows a
comparison of the integrated damage energy versus primary atom
recoil energy for nickel irradiated with REF, NSF, fission,
degraded fission (CP-5 reactor), and 14 MeV neutrons. It is clear
from this figure that the REF damage distribution is quite similar
to that obtained by steady-state reactor irradiation.

All irradiations in the present experiments were performed at
5 K in the REF vertical thimble VT-2. The samples were in a

helium exchange gas during irradiation and annealing. With the
exception of the samples used in the DLTS experiments, all samples
were degenerate or nearly so. Table 1 lists the resistivity of

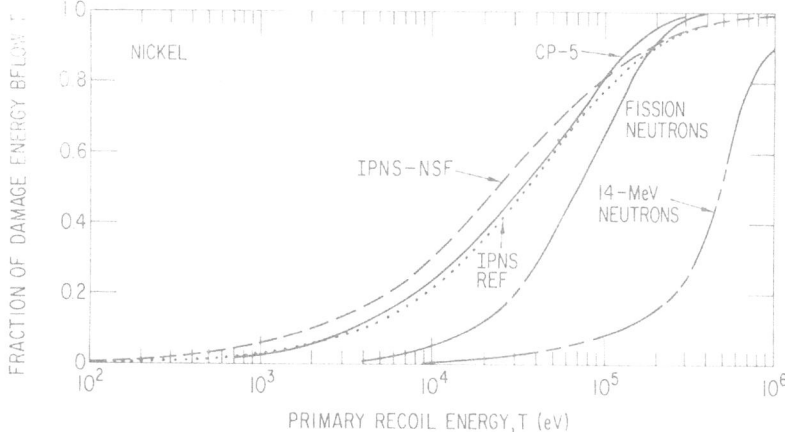

Fig. 3. Comparison of integral damage-energy distributions in
nickel irradiated with NSF, REF, degraded fission (CP-5),
fission and 14 MeV neutrons.

each sample at four temperatures before irradiation as well as the
carrier concentration at 5 K before irradiation. Initial carrier
concentrations were determined from Hall-effect measurements
either at room temperature or liquid helium temperature for all
the compound semiconductor samples, but were estimated from known
room temperature carrier concentration vs. resistivity curves for
the silicon samples. Contacts were attached to sliced samples
using ultrasonic indium soldering in all cases. All silicon
samples were grown by the Czochralski technique and contain a high
concentration of oxygen estimated to be 10^{17} to 10^{18} cm^{-3}.

The base to collector of a 2N2270 (NPN) transistor was used
as the p$^+$n diode for all DLTS experiments. It was estimated that
the dopant concentration in the n-type collector was in the low
10^{16} cm^{-3} range from breakdown measurements. Although the dopant
for the collector is not known, it is most likely to be either
phosphourus or arsenic. It is also most likely that this material
was grown by the Czochralski (CZ) technique. This is confirmed by
a strong A-center (vacancy/oxygen) energy level at 0.16-0.17 eV
following irradiation and annealing to above the vacancy migration
temperature range.

CARRIER REMOVAL VS. FLUENCE AT 5 K

Figures 4 and 5 show a decrease in normalized conductivity at 5 K with increasing proton fluence on the REF target. Conductivities have been normalized to the before-irradiation values

Table 1. Initial resistivity before irradiation as a function of temperature. Carrier concentrations at 5 K before irradiation are also given.

Material (Dopant)	RESISTIVITY (Ω-cm)				n_o (cm^{-3})
	Room T	77 K	21 K	5 K	5 K
Si(Sb)-Cz	3.85×10^{-3}	2.53×10^{-3}	2.15×10^{-3}	2.09×10^{-3}	1.8×10^{19}
Si(P)-Cz	1.28×10^{-2}	1.36×10^{-2}	1.02×10^{-2}	0.966×10^{-2}	3.7×10^{18}
Si(As)-Cz	1.29×10^{-2}	3.76×10^{-2}	19.7×10^{-2}	41.4×10^{-2}	?
Si(B)-Cz	1.57×10^{-2}	3.60×10^{-2}	3.41×10^{-2}	3.20×10^{-2}	7.0×10^{18}
CdTe (In)	1.33×10^{-2}	1.53×10^{-2}	2.05×10^{-2}	2.07×10^{-2}	7.8×10^{17}
CdS (Cl)	1.96×10^{-3}	1.57×10^{-3}	1.65×10^{-3}	1.65×10^{-3}	1.3×10^{19}
GaAs (Te)	1.16×10^{-3}	1.08×10^{-3}	1.11×10^{-3}	1.11×10^{-3}	2.0×10^{18}

obtained from Table 1. Proton fluence (number of protons hitting the uranium target) can be converted to fast neutron fluence (n/cm^2) at the sample in the vertical thimble by multiplying by the factor 2.0×10^{-2} n/cm^2/p which was obtained by integration of the neutron spectrum in Fig. 2 from the lowest energy which will produce displacements up to the maximum energy measured.

An average of eleven DC conductivity measurements, taken in rapid succession with a digital voltmeter, data scanner and computer, was used to obtain each observed data point. These averages were obtained and plotted at intervals of about 2.5×10^{16} proton fluence so as to form a nearly continuous plot. Individual data points are not therefore shown on these two figures. All conductivity measurements were made in the dark. No conductivity changes were observed when the samples were initially inserted into the radiation position suggesting that photoconductivity from induced radioactity in the cryostat was small at the beginning of the experiment. The accelerator beam was interrupted several times during the run and measurements taken to check for gamma-induced minority-carrier injection and trapping. Conductivity changes under these conditions were also negligible. Data for the conductivity vs. fluence of Si(As) is

not shown since there was some apparent difficulty with sample
contacts or sample heating during the first half of the irradiation for this sample.

It can be seen from Figs. 4 and 5 that the effect of the
incremental irradiation damage on the conductivity decreases with
increasing fluence for all samples regardless of the material or
its initial doping level. There is a definite trend toward

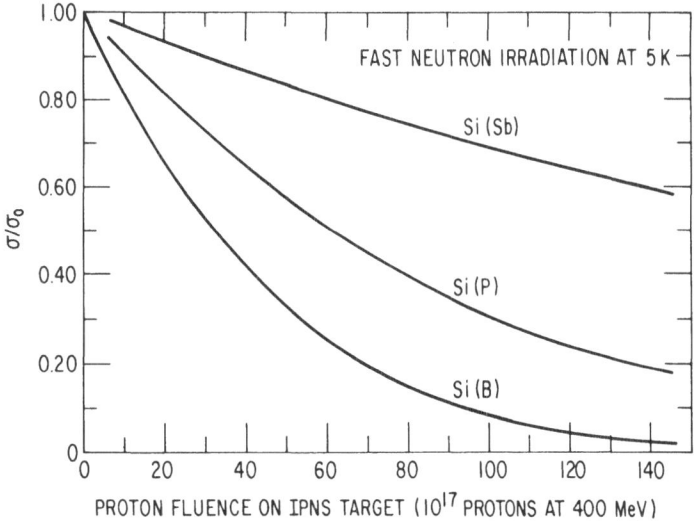

Fig. 4.   Fractional decrease in electrical conductivity vs.
               proton target fluence for degenerate silicon doped with
               Sb, P and B at 5 K.

saturation. This is illustrated by plotting the slopes of Figs. 4
and 5 $d(\sigma/\sigma_0)/d\phi$ vs. normalized conductivity $\sigma(\phi)/\sigma_0$ as shown in
Fig. 6. Each point of Fig. 6 represents the result of a least
squares polynomial fit through 5 data points of σ/σ_0 vs ϕ. The
surprising fact emerging from Fig. 6 is that, to a good approximation, all data can be approximated fairly well by the following
linear relationship:

$$d(\sigma/\sigma_0)/d\phi = -K(\sigma/\sigma_0) \tag{1}$$

where ϕ is the fluence and K is a proportionality constant
representing the slope of the approximating straight line. Also,

at $\sigma = \sigma_0$ and $\phi=0$, K is given by the $\sigma/\sigma_0 = 1$ intercept. Using these limits, Eq. (1) can be integrated to yield the simple expression:

$$\sigma(\phi) = \sigma_0 e^{-K\phi} . \tag{2}$$

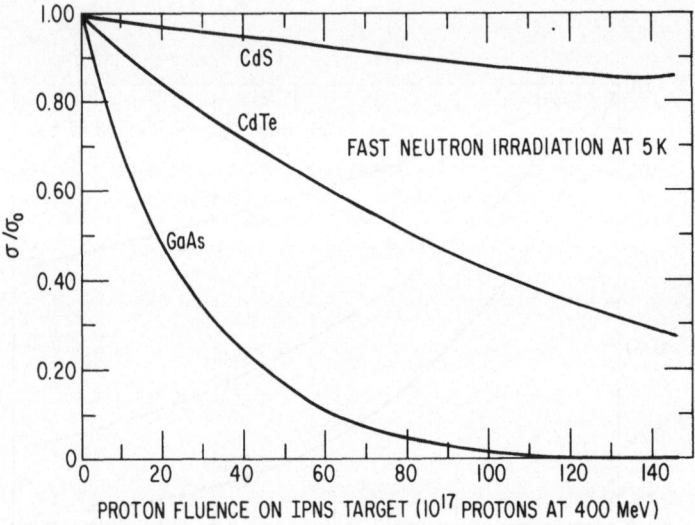

Fig. 5. Fractional decrease in electrical conductivity vs. proton target fluence for degenerate CdS, CdTe and GaAs at 5 K.

The values of K for each sample are listed in Table 2 and are expressed in terms of proton fluence on the uranium target, Kp, as well as neutron fluence on the sample, K. The initial rate of change of conductivity near $\phi=0$ can be expressed as a carrier removal rate by trapping under the assumption that the rate at which carries are removed with fluence is large compared to the rate at which carrier mobility changes with fluence. This is considered a good approximation in the present experiments since for these degenerate samples, the mobility at low temperature should be dominated by ionized-impurity scattering from the high concentration of ionized donors or acceptors. Furthermore, a decrease in mobility with fluence, induced from charged defect centers, should cause a sublinear decrease in conductivity. In fact, just the opposite effect is observed in Figs. 4 and 5. The initial carrier removal rate (i.e., at $\phi=0$), under the previous assumption, can be obtained from the expression:

$$dn/d\phi \bigg|_{\phi=o} = n_o d(\sigma/\sigma_o)/d\phi \bigg|_{\phi=o} = -n_o K \qquad (3)$$

using the values for n_o listed in Table 1. These initial carrier removal rates are also listed in Table 2. Normalized conduc-

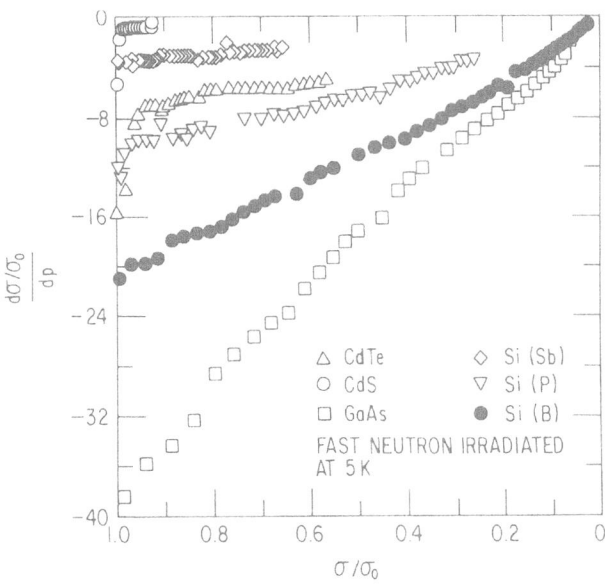

Fig. 6.   The normalized rate of conductivity change $d(\sigma/\sigma_o)/d\phi_p$ vs. fractional normalized conductivity (σ/σ_o) for all semiconductor samples at 5 K.

Table 2.   Conductivity damage constant K and initial carrier removal rates.

Material	n_o (cm^{-3})	K_p	K (cm^{+2})	$-dn/d\phi_p$ (cm^{-3})	$-dn/d\phi_n$ (cm^{-1})
Si(Sb)	1.6×10^{19}	3.7×10^{-20}	1.85×10^{-18}	0.59	29.5
Si(P)	3.7×10^{18}	12.0×10^{-20}	6.0×10^{-18}	0.44	22.2
Si(B)	7.0×10^{18}	21.4×10^{-20}	10.7×10^{-18}	1.50	75.0
CdTe	7.8×10^{17}	8.0×10^{-20}	4.0×10^{-18}	0.062	3.1
CdS	1.3×10^{19}	1.0×10^{-20}	$50. \times 10^{-18}$	0.13	6.5
GaAs	2.0×10^{18}	39.0×10^{-20}	$18. \times 10^{-18}$	0.78	39.0

Table 3.   Damage rate, carrier removal rate and ratio for 5K fast neutron irradiations.

Material	Calculated Damage Rate (cm^{-1})	Observed Removal Rate (cm^{-1})	$\dfrac{\text{Removal}}{\text{Damage rate}}$
CdTe	7.7	3.1	0.4
CdS	22.9	6.5	0.3
GaAs	17.9	39.0	2.2
n-Si	50	25.0	0.5
p-Si	50	75.0	1.5

tivity constants are given for proton fluence $[K_p = -d(\sigma/\sigma_o)/d\phi_p]$ and neutron fluence $[K = -d(\sigma/\sigma_o)/d\phi]$ in Table 2. Conversion from proton fluence on target to neutron fluence on sample is obtained by dividing by 2.0×10^{-2} $n/cm^2/p$ as explained previously.

The expected damage rate in silicon, calculated from the spectrum averaged damage energy cross section $\langle\sigma_d T\rangle$, is obtained from the expression

$$dN_D/d\phi = N_T \langle\sigma_d T\rangle/2E_d \qquad\qquad (4)$$

where N_D is the concentration of displacements, ϕ the neutron fluence, $N_T = 5 \times 10^{22}$ cm^{-3}, the concentration of silicon atoms in the lattice, and $E_d = 25$ eV, the displacement threshold. The spectrum averaged damage-energy cross section for silicon, $\langle\sigma_d T\rangle =$

50 keV-barns, was obtained using the program DISCS.[8] This calculation predicts a damage rate of about 50 displacements/cm^3 per n/cm^2. It should be noted that the observed carrier-removal rates (25 cm^{-1} for n- type and 75 cm^{-1} for p-type) bracket this calculated damage rate of 50 cm^{-1}. The damage rates for the other semiconductors has not yet been calculated, however, we can make a rough estimate based on the Kinchin-Pease[9] model and scaling to the calculated silicon damage rate using:

$$\frac{dN_D/d\phi)_{compound}}{dN_D/d\phi)_{Si}} = \frac{\bar{N}_T \bar{\Lambda})_{compound}}{\beta N_T \Lambda)_{Si}} \tag{5}$$

where:

$$\Lambda = \frac{4 M_n M}{(M_n + M)^2}$$

and where M_n is the neutron mass, M the mass of the displaced atom, and $\bar{\Lambda}$ is the average for the compound, N_T the concentration of target atoms in the lattice and β the ionized loss factor which we have taken to be about 0.5 for silicon and 1.0 for the compounds. A comparison between the carrier-removal rates and the estimated damage rates is shown in Table 3.

It is seen that the observed removal rates are within a factor of about two of the calculated damage rates in all cases. In general, some spontaneous recombination of Frenkel pairs in the cascade is expected to lead to a ratio less than one if it is assumed that one carrier is removed per displacement. It is known, however, that many defects in semiconductors contain multiple energy levels in the band gap and the possibility exists for the above ratio to be perhaps as high as three or four, for multiple carrier trapping. Under the circumstances, therefore, the initial carrier removal rates appear to fall in the range of values expected from displacement theory. It should also be mentioned that errors as large as a factor of two might result from estimating the actual carrier concentration from Hall measurements.

We have assumed above that a displacement collision produces an array of isolated point defects in the cascade and that the carrier-removal rate is some measure of the point defect concentration and carrier trapping efficiency. An alternate model exists, however, which has been very successful in explaining a number of experimental facts in non-degenerate neutron irradiated semiconductors.

This "cluster model" by Gossick[10] and by Crawford and Cleland[11] is based on the fact that the carrier removal is so great within the cascade compared to the surrounding undamaged crystal, that locally the Fermi level moved closer to midgap. The result of this is band bending and the formation of a depletion region around the cascade whose radius may be roughly estimated from the expression[10,11]

$$V = 4\pi r_2^3/3 = 4\pi\varepsilon\psi_p r_1/qn_o = \frac{\ln(1-f)}{\Sigma_v\phi} \tag{6}$$

where V is the volume associated with the space charge, ε is the dielectric constant, Ψ_p the difference in Fermi energy between the cluster and the surrounding matrix, q is the electronic charge, n_o the carrier concentration in the undamaged region, r_1 and r_2 the radii of the disordered region and space charge regions, f is the fraction of crystal occupied by insulating spheres of radius r_2, $\Sigma_v = N_T\sigma_s$, the probability per cm that a fast neutron will produce such a region, and ϕ the fluence. The conductivity for a conducting medium containing insulating spheres is then given by:[12]

$$\sigma = \sigma_o \frac{1-f}{1 + \frac{1}{2}f} \cdot \tag{7}$$

Eqs. (6) and (7) then lead to the following conductivity fluence dependence:

$$\sigma = \sigma_o \frac{e^{-V\Sigma_v\phi}}{\frac{3}{2} - \frac{1}{2}e^{-V\Sigma_v\phi}} \tag{8}$$

which is similar to but not identical with Eq. (2).

The cluster model predicts that the carrier-removal rate should be independent of initial carrier concentration near $\phi=0$.[11] Stein and Gereth, however, have observed that for all types of silicon irradiated with fission neutrons at room temperature, the removal rate increases slightly with carrier concentration.[13] Furthermore, the removal rates observed by Stein[3,4] in n- and p-type silicon irradiated at 76 K with fission neutrons fit reasonably well with the data he obtained for room temperature irradiations.[13] If we extrapolate Stein's room temperature removal rate curves to the high carrier concentrations used in the present experiments, we would predict removal rates of 30, 25 and 50 cm^{-1} for Si(Sb), Si(As) and Si(B) respectively as compared to the observed values of 29.5, 22.2 and 75 cm^{-1}. The

present data agrees, therefore, to within the experimental error,
with the extrapolation of Stein's curves of room temperature
data. We conclude, therefore, that the carrier-removal rate for
neutron irradiated silicon is relatively independent of
temperature from 5 to 300 K in contrast to electron-irradiation
removal rates which vary by orders of magnitude over the same
temperature range. Since many point defects are known to anneal
in silicon below room temperature, the temperature independence of
carrier-removal rate strongly suggests that the basic premise of
the cluster model is correct. Any annealing observed in low-
temperature irradiated material should, therefore, represent a
modulation of the depletion radius r_2. It should be noted,
however, that highly degenerate semiconductors cannot support an
appreciable depletion region.[10] We expect therefore that $r_1 \simeq r_2$
initially in our samples.

We can calculate the conductivity damage constant K from
cluster theory and compare the results to the observed values in
Table 2. The mean recoil energy for silicon atoms in our
experiment is about 47 keV. We therefore expect a mean range of
about 70 nm, or a cluster radius of about 35 nm. This leads to a
cluster volume of 1.8×10^{-16} cm^3. From the derivative of Eq. (8),
we find that:

$$-\frac{d(\sigma/\sigma_o)}{d\phi}\Bigg]_{\phi=o} = \frac{3}{2} V\Sigma_v = K_{cluster} \quad . \tag{9}$$

using $\Sigma_v = N_T\sigma_s = (5 \times 10^{22} cm^{-3})(3 \times 10^{-24} cm^2)$, we calculate that
$K = 40.5 \times 10^{-18}$ cm^2 compared to experimental values from Table 2 of
2 to 10×10^{-18} cm. We see that the calculated conductivity damage
rate is a factor of 4 times the observed value for p-type silicon
and 20 times the observed value for n-type Si.

It appears that the temperature independence of the initial
damage rate supports the cluster model in general but that the
dependence of removal rate on carrier concentration, the
conductivity damage constant and the form of Eq. (8) do not in
detail.

Finally, we must contend with the fluence dependence of the
conductivity represented by Eq. (2). Although not generally
recognized, the exponential decay of conductivity or carrier
concentration with fluence is a rather universal phenomenon of
radiation damage to semiconductors. As we have seen in Fig. 6,
this effect appears to be independent of material or initial
carrier concentration. In fact, at least one and perhaps several
samples were driven through the Mott transition from degenerate to

non-degenerate with no apparent change in the behavior of Eq.
(2). This same relationship has also been observed for strongly
compensated CdTe (n = 1.6×10^{13} cm^{-3}) electron irradiated at
4.2 K[15] and room-temperature neutron-irradiated Ge.[16] There are
in fact, indications of this effect whenever the defect
concentration becomes an appreciable fraction of the free carrier
concentration.

This effect is believed to be caused by the saturation of
free carriers available for measurement. If F is the fraction of
initial free carrier concentration removed by a fluence ϕ, then
f = 1-F $\cong$ σ/σ_0 is the fraction of carrier concentration not
removed. For a small increase dN_D in the defect concentration
(assuming one carrier removed per defect), dF must increase in
proportion to the fraction of carrier concentration still
available as well as the fraction of carriers removed dN_D/n_0,
therefore,

$$dF = (1 - F) \frac{dN_D}{n_o} . \tag{10}$$

Integration of this equation, using F→1 as N_D→∞ and F = 0 for N_D =
0, yields

$$f = 1 - F = \frac{\sigma}{\sigma_o} = \exp (-N_D/n_o). \tag{11}$$

Since $N_D/n_o = (N_T \sigma_s \bar{\nu} /n_o)\phi$ using a simple Kinchen-Pease

Table 4.   Comparison of experimental conductivity damage
            constant $K_{exp} = - d(\sigma/\sigma_o)/d\phi$ with constants calculated
            from the "Available Carrier Model" ($K_{carrier}$) and the
            "Cluster Model" ($K_{cluster}$).

Material	$K_{exp}(cm^2)$	$K_{carrier}$	$K_{cluster}$
Si(Sb)	1.85×10^{-18}	3.1×10^{-18}	40.5×10^{-18}
Si(P)	6.0×10^{-18}	13.5×10^{18}	40.5×10^{-18}
Si(B)	10.8×10^{-18}	7.1×10^{-18}	40.5×10^{-18}
Ge(ORNL)*	1.3×10^{-17}	1.8×10^{-17}	3.5×10^{-19}

*Data from Ref. 16

expression for the number of defects,[9] comparison of this equation
with Eq. (2) suggests that the conductivity damage constant K
should be calculated from

$$K_{carrier} = N_T \sigma_s \bar{\nu} / n_o .\qquad\qquad(12)$$

This factor should be multiplied by a factor representing the
number of carriers removed per defect, however, since this number
is not known, we have taken this additional factor to be unity.
Table 4 compares the observed conductivity damage rate constants
with those based on our "available carrier model" and the "cluster
model".

We have also analyzed room temperature neutron data for
Germanium ($n_o \sim 1 \times 10^{18}$ cm^{-3}) using the flux spectrum for the ORNL
graphite reactor[1] and a simple Kinchen-Pease model. The results
of these calculations are also shown in Table 4 for the cluster
and carrier models. These calculations for Germanium are based on
our estimates $\bar{E}_R$ = 13.4 KeV ,$V = 1.8 \times 10^{-18}$cm^3 and N_D / ϕ = 17.9 cm^{-1}.
It appears that degenerate Germanium is an even better test to
compare these two models.

It can be concluded, on the basis of the previous experiments
and calculations, that the cluster model explains the apparent
lack of temperature dependence on damage. However, it does not
explain the fluence dependence for degenerate semiconductors. It
appears that the agreement between the calculated damage rate and
the carrier removal rate is reasonably good for the materials
investigated, however, this should be subject to further
experimentation over a wider range of carrier concentrations.

ISOCHRONAL ANNEALING

Samples were isochronally annealed (for 5-min intervals)
following irradiation. Annealing data are shown for the silicon
samples in Fig. 7, for the II-IV compounds in Fig. 8 and for the
GaAs samples in Fig. 9. The data points shown in Figs. 8 and 9
are the average of 24 individual readings per point. Data points
are not shown for the silicon samples in Fig. 7 for clarity,
however, they were taken after the same annealing schedules as the
other samples. All measurements were made at 5 K. The data have
been normalized to the resistivities measured after irradiation,
but before any annealing.

Resistivities were observed to systematically change a small
amount following irradiation, but before the first anneal,
indicating the presence of some radioactive effects, minority

carrier trapping or sample cooling. However, these changes were generally smaller than 30% and data was not taken until the drift in the resistivity was smaller than 1/2%. Annealing was interrupted for 8 hours after the 90 K annealing step. Again small drifts in the resistivity of the compound semiconductors was observed, however, subsequent annealing to 100 K caused no abrupt changes in the annealing curves. This effect suggests the presence of radioactive-decay induced minority carrier trapping to some extent in the compound semiconductors. The general upward drift of the CdS sample can also perhaps be interpreted as due to radioactive decay. We shall discuss the annealing of these samples by groups in the following subsections.

Annealing of The Silicon Samples

Extensive experimentation by others have lead to a wealth of annealing results from which to draw comparisons. Table 5 presents a summary of known defect annealing stages taken from several sources.[17-20] Since the position of the annealing stage varies considerably with annealing time and defect concentration, the following table should be considered suggestive but not exhaustive. It is clear that some confusion still exists concerning the annealing temperatures of V^0 and $V^=$.[21]

Several small reverse annealing stages are observed in n-type silicon (Fig. 7) at 22, 33 and 41 K for Si(Sb), Si(P) and Si(As) respectively. In addition the Si(P) and Si(As) also show annealing in the range of 40 to 50 K. The mechanism of reverse annealing at the two lowest temperature stages are unknown, however, the 40-50 K reverse annealing agrees with the annealing of the (V-O) $^*_{\overline{100}}$ a metastable state of the vacancy-oxygen A-center with the oxygen in a ⟨100⟩ orientation. Additional reverse annealing at 64 to 72 K appears in all the n-type samples and is probably associated with the migration of the isolated vacancy in the negative charge state. Annealing in the range of 82 to 96 K may involve the annealing of the (V-O) $^*_{\overline{1}11}$ center or neutral vacancy migration. A large reverse annealing at about 140 K has been attributed by Stein to thermally stimulated carrier emission, not defect annealing.[3] In fact all the reverse-annealing peaks up to this peak could be due to the thermal release of trapped carriers.

The recovery from 140-200 K appears to be due to the migration of the vacancy in the double negative ($V^=$) charge state in n-type material and the neutral charge state (V^0) in p-type material. There is some indication that silicon interstitials also migrate in this range.[19] The annealing stage in p-type Si(B) between 60 to 120 K is due to the positive charge state vacancy

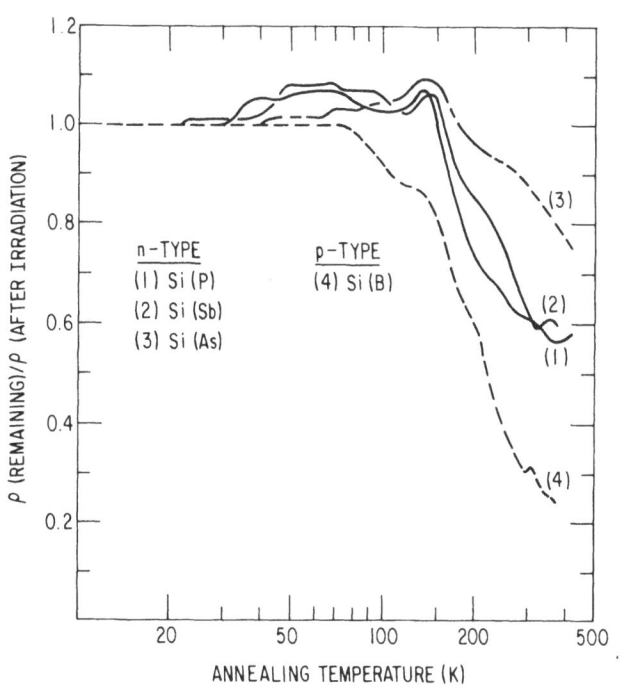

Fig. 7.   Isochronal annealing (5 min.) for silicon doped with
P, Sb, As, and B. The resistivity after each
isochronal annealing step, normalized to the
resistivity after irradiation, is plotted as a
function of annealing temperature.

Table 5. Approximate annealing temperatures for various defects in silicon taken from Refs. 17–20. The temperature at which the defect disappears is listed unless otherwise mentioned.

Defect	Corbett Ref. 17	Stein Ref. 18	Watkins Ref. 19	Kimmerling Ref. 20
V^+				60–140 K
V^0	60–80 K	150 K	150 K	150–180 K
V^-				60–70 K
$V^=$	170–200 K	70 K		90–120 K
$(V-O)^{*-}_{100}$		78 K	45 K	120 K
$(V-O)^*_{111}$	in 40 K		in 70 K	in 80 K
	out 100 K		out 90 K	out 120 K
$(V-O)^-$		600 K		in 230 K
$V-B_s$	280–340 K			
$V-P_s$	380–420 K	400 K		
$V-As_s$	420–460 K	440 K		
$V-Sb_s$	440–490 K	460 K		
Si_I^-(G25)	150–180 K			140 K
B_I	280–350 K			
Vacancy Clusters (n- irrad.)	200–500 K			

(V^+) migration. Stein has suggested that annealing from 200–350 K is the result of vacancy loss from clusters in both n- and p-type neutron irradiated material.[18] The A-center (V-O) is observed to increase in this same temperature range suggesting the liberation of vacancies from clusters into the lattice.[18] Little evidence is seen in our data for dopant interstitial or E-center (vacancy – substitutional impurity) annealing.

Annealing of the II-VI Specimens

The annealing of CdS and CdTe is shown in Fig. 8. Reverse annealing dominates in both samples up to about 400 K. Several stages are observed, however, and can be identified.

The understanding of defects in the II-VI compounds has greatly increased in recent years from Electron Spin Resonance (EPR), luminescence and stoichiometry experiments.[22] These compounds all have two distinct displacement thresholds which are well separated in fast electron irradiation energy provided the masses of the anion and cation atoms differ appreciably. Radiation damage produced by fast electrons can then be confined to a single sublattice by irradiation between the two thresholds thereby making assignments of annealing stages to defects on a particular sublattice a certainty.[23] Furthermore, EPR experiments

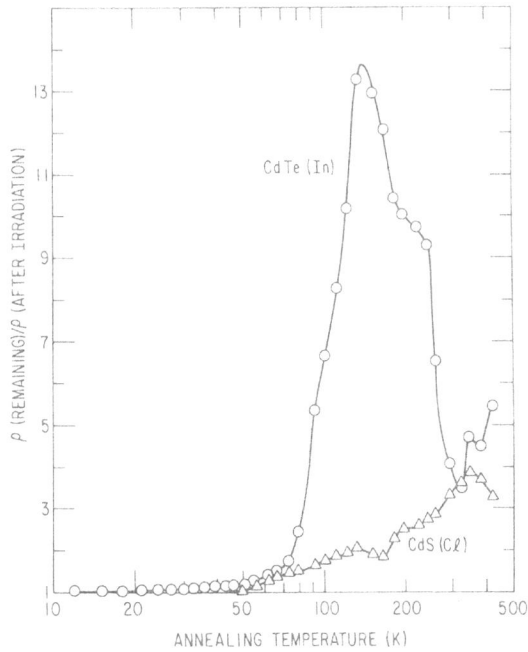

Fig. 8.   Isochronal annealing (5 min.) for CdTe doped with In
and CdS doped with Cl. Normalized resistivities
similar to Fig. 7 are plotted vs. annealing
temperature.

indicate no vacancy migration on either sublattice below room temperature.[22] Any annealing stages below room temperature must, therefore, be confined to interstitial migration, defect reorientation, or carrier emission and trapping. The metal sublattice vacancies are known to be deep acceptors while the anion vacancies are shallow donors.[22,23] The interstitial charge states are more uncertain. However, difficulty in observing these defects with EPR directly[22] suggests that they may be neutral on the metal sublattice.

Watkins has shown that Frenkel pairs exist in several configurations on the metal sublattice in ZnSe and that they reorient into different configurations in the range of 60 to 180 K in three or four annealing stages.[22] Vook has observed an increase in anisotropic thermal conductivity from 60 to 150 K in electron irradiated CdS followed by a loss of this anisotropy in a stage at 180 K.[24] This is followed by nearly complete recovery in the range of 210 to 270 K. These same features are evident in Fig. 8. The anisotropic orientable defects in Vook's work on CdS are quite suggestive of the various Frenkel pair annealing stages observed in ZnSe by Watkins with EPR; furthermore, the annealing temperatures are nearly the same.

Elsby and Meese[23] have shown, using electron irradiations at energies between the sulfur and the cadmium displacement thresholds and using luminescence measurements that the 180 K stage is associated with annealing of sulfur sublattice damage while the annealing from 210 K to above room temperature occurs only after irradiation above the cadimium threshold. This leads to a very simple picture of the annealing of neutron irradiated CdS.

Frenkel pairs in the cluster reorient in the range of 50 to 130 K with some occasional recombination of defect pairs. The general rise in resistivity suggests that more donors are lost than acceptors. If the vacancies are the primary electrically active defects, this implies the loss of more sulfur vacancies (donors) than cadmium vacancies (acceptors). This is reasonable in CdS since sulfur is the lighter of the two ions and hence is expected to be displaced in larger numbers than cadmium.

Because vacancies are not mobile below room temperature, the stage between 160 to 200 K must most likely be due to long range migration of sulfur interstitials. Since some of the sulfur vacancies are known to survive at room temperature[22], more interstitials (neutral or acceptors) are lost than vacancies (donors). We would therefore expect a decrease in resistivity in this range. In fact, just the opposite is observed From 200 up to 350 K, long range cadmium interstitial migration occurs.[23] It

would therefore be expected that the annihilation of cadmium
vacancies (acceptors) with cadmium interstitials (neutral or
donors) should produce a decrease in resistivity in this stage.
In fact, this has been observed in electron irradiated CdS[25] and
again differs from our result. Finally, the annealing from 250 K
to our highest anneal temperature has been identified as a sulfur
vacancy migration stage.[26] This implies a loss of shallow donors
and a resistivity increase again contrary to our results for
neutron irradiation and observations in electron irradiated
material.[26] These resistivity changes in the wrong direction
suggest additional defects are present in the neutron irradiations
which were not present following electron irradiation which are
interacting with the migrating species in some unknown way.

The annealing of CdTe is dominated by a larger reverse
annealing peak whose size reflects the lighter doping level of
this sample compared to the CdS. Almost identical reverse
annealing of very lightly doped (8.1×10^{14} cm^{-3}) n-type CdTe has
been observed by Caillot while monitoring photo-conductivity.[27]
An inspection of Fig. 8 shows that CdTe anneals slightly in the 30
to 70 K temperature range followed by a large reverse anneal from
70 to 140 K. Normal annealing behavior is then observed in stages
from 140 to 210 K and from 210 to 320 K. Reverse annealing is
again seen from 320 to 420 K.

Because CdTe melts at a much lower temperature than CdS,
corresponding annealing stages should occur at somewhat lower
temperatures. On this basis, we tentatively assign the annealing
in the range of 30 to 70 K to Frenkel-pair reorientations. The
annealing from 70 to 140 K has been conclusively associated with
Tellurium sublattice damage by Bryant, Webster and Cox from
threshold measurements.[28,29] By analogy to CdS it is most likely
that this recovery is caused by long range motion of Tellurium
interstitials. Tagushi and Inuishi have determined a migration
energy of 0.2 eV for this stage based on the intensity of the 1.1
eV luminescence band.[30] If the Tellurium vacancy is the origin of
the 1.1 eV band as proposed by Bryant, then long range Tellurium
interstitial migration is expected to reduce the intensity as
observed.[28] It should be noted that the resistivity change is
again in an unexpected direction suggesting large concentrations
of other defects not observed in electron irradiated material.

Tagushi and Inuishi have argued that the recovery from 320 to
420 K is the annealing of an acceptor at $E_v + 0.15$ eV.[30] They
assign this stage to the migration of cadmium vacancies based on
the similarity of their annealing kinetics [1st order,
$\nu = (10^{15}$ s^{-1}) exp $(-0.85$ eV/kT)] with those observed for other
metal ion vacancies in the II-VI system. Again we see the
resistivity changing in an unexpected direction.

The annealing behavior from 140 to 210 K is assigned to cadmium interstitial long range migration. This assignment places this motion at a lower annealing temperature in CdTe than CdS which is reasonable considering the melting temperatures. This would leave the 210 to 320 K annealing to Tellurium vacancy migration. Again, the temperature is somewhat lower than sulfur vacancy migration in CdS. If Tellurium vacancy migration occurs in this temperature range, this would be the only known vacancy in a II-VI compound to migrate below room temperature. It seems, however, that known annealing stage assignments are forcing this conclusion from either end. This conclusion suggests that a search for the Tellurium vacancy in CdTe be investigated using lower than room temperature irradiations.

Annealing of the GaAs Specimens

The isochronal annealing of GaAs is shown in Fig. 9. It appears that the GaAs annealing has characteristics between those of the II-VI compounds and those of silicon. A rather large reverse annealing is observed from 18 to 80 K which contains considerable structure. This is followed by normal annealing behavior from 80 to 260 K, a small reverse anneal from 260 to 320 K and a normal annealing behavior from 320 to 400 K.

The annealing of radiation damage introduced into GaAs by a variety of incident particles has been reviewed by Lang.[31] The annealing for electron irradiation at 77 K can be generally characterized as consisting of three first order stages centered approximately at 235, 280 and 500 K.[32] Reverse annealing between 30 to 80 K has been observed in n-GaAs (n_o=1x10^{16}cm^{-3}) irradiated with 2x10^{13}e/cm^2 (0.9 MeV) which has been attributed to reversible electronic processes.[32]

The annealing following 77 K high energy electron (5 to 30 MeV), or fast neutron irradiations appears to be quite different from 1 MeV electron irradiation. Broad annealing from 77 K to room temperature has been observed by Stein[33] for neutron irradiations and by Kalma, et al.[34] for high energy electron irradiations.[35] Picraux has found using channeling studies of disorder following 77 K ion inplantation, that the disorder on both sublattices in GaAs appears to anneal at about the same rate over this temperature range.[35] It appears from these studies that the annealing presently observed in the temperature range from 80 to 300 K might be dominated by cluster annealing and is distinctly different from the point defect annealing observed following 1 MeV electron irradiation. Since little is known about the microscopic details of defects in GaAs at present,[31] further interpretation must be postponed until additional low temperature experiments have been performed.

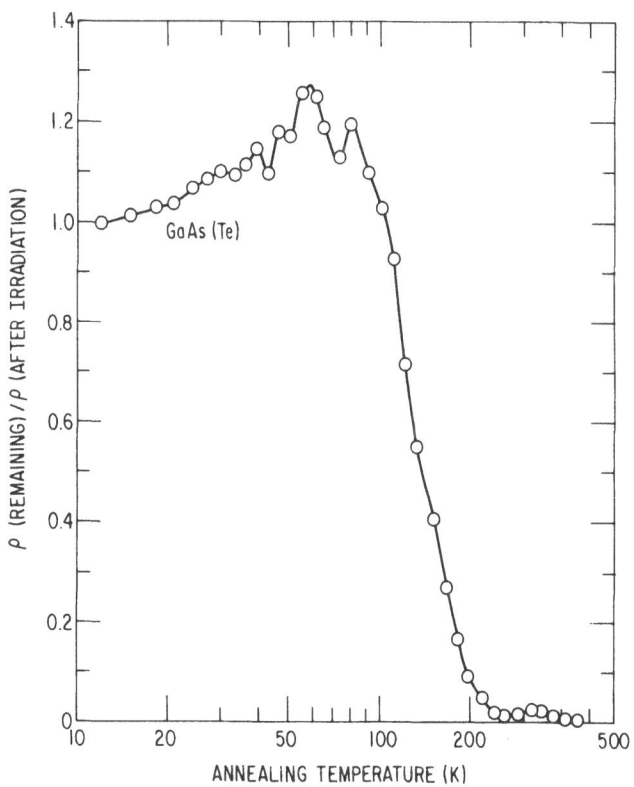

Fig. 9.   Isochronal annealing (5 min.) for GaAs doped with
            Tellurium.

DLTS EXPERIMENTS ON NEUTRON IRRADIATED n-TYPE SILICON

Very few DLTS experiments have been performed on any material
irradiated below room temperature. Because of defect annealing
during the measurement, only shallow states can be observed at the
lower isochronal annealing temperatures. Long time-constant
capacitance-transient measurements ($\tau \approx 100s$) allow observation of
states only as deep as 0.25 eV at 100 K.[36] After annealing to

higher temperatures, deeper levels can be observed. In spite of these limitations, Kimerling, et al. have observed a number of deep levels introduced into silicon by proton irradiation at 45 K.[36] In particular, they observed four shallow defect levels which disappear in the range of 80 to 120 K. The energy of only one of these levels (E2=0.09 eV) has been determined and identified as the isolated vacancy, however, two other levels, E3 and E3', have been associated with the $(V-O)^*_{100}$ and $(V-O)^*_{111}$ defects respectively.[36] From our isochronal annealing results of the previous section, it is expected that these levels would also be seen in neutron-irradiated material. In fact, none of the above low-temperature DLTS peaks has been observed in the present experiments. Several other interesting results have been obtained from these experiments, however.

In the present measurement a test diode is placed in the cryogenic radiation effects facility (REF) and a test DLTS trace is observed as the specimen is brought to room temperature after remaining at 5 K for about 24 hours. During this period, protons were incident on the neutron scattering facility (NSF) target which is shielded from the REF target by about 2.5 m of iron and a relatively smaller thickness (0.1 m) of lead. Since no intentional flux of protons was incident on the REF target, it was surprising to observe levels which could be associated with room-temperature neutron irradiation (top trace of Fig. 10) after heating to room temperature. Figure 10 shows the growth of these four levels, E_0 (0.10 eV), E_1 (0.16 eV), E_2 (0.23 eV), and E_3 (0.40 eV) as a function of proton fluence on the NSF target. The production of these levels, following room temperature annealing, is shown in Fig. 11. Since three of the above levels had been reported previously following room-temperature neutron irradiation,[37] there seemed little doubt that neutrons from the NSF target were irradiating the sample through the iron and lead shielding.

It is well known that the total neutron cross section for iron has a resonance window at 24 keV. By assuming that all neutrons hitting the sample were 24 keV neutrons, and using a Kinchen-Pease damage model,[9] it was estimated that a flux of 24 keV neutrons of about 10^6 n/cm^2/s would be needed to produce the observed damage rate. This flux can be estimated from the known spectrum of neutrons measured near the NSF target.[6] After estimating the fraction of these neutrons which occurred in a 2 keV window at 24 keV and scaling for the additional distance to the REF target, a flux of about 10^6 n/cm^2/s is again found. It therefore appears that a small leakage flux between the two targets has been observed. Since the defect concentration after a week of irradiation was about 5×10^{13} cm^{-3}, this leakage flux should be negligible for most experiments, however, these results do demonstrate the extreme sensitivity that can be achieved in DLTS measurements (0.1 ppb).

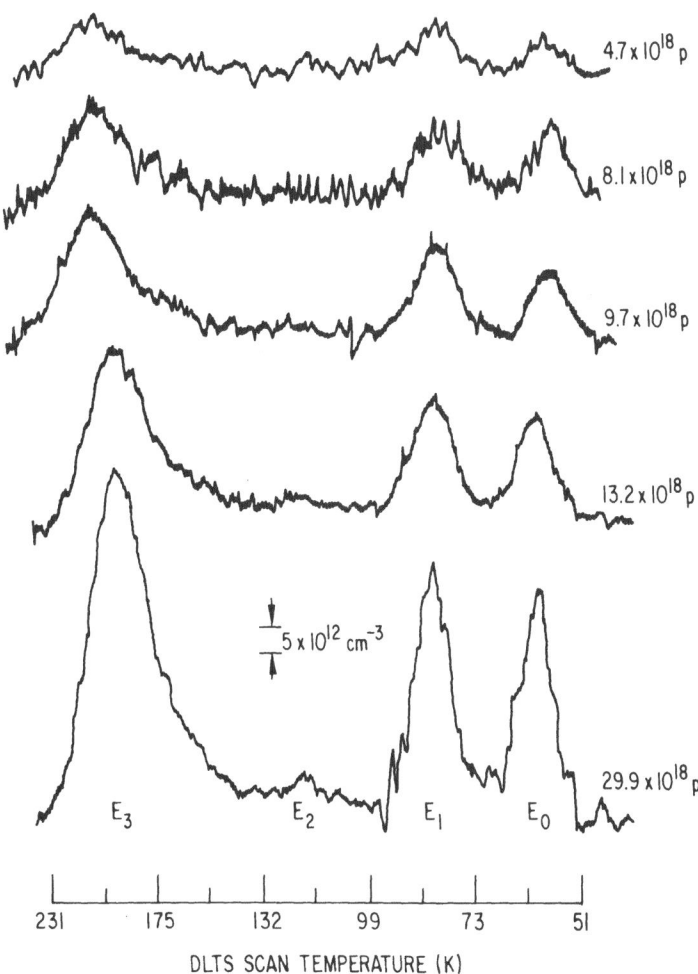

Fig. 10. DLTS majority carrier spectra obtained for a p⁺n junction diode in the radiation effects facility (REF) Proton fluences on the neutron scattering facility (NSF) as are shown to the right of each scan. Each DLTS scan was obtained after an anneal to room temperature. A trap concentration scale of 5×10^{12} cm⁻³ is indicated to calibrate the vertical axis. Energy level values of the E_o, E_1, E_2, and E_3 peaks are given in the text.

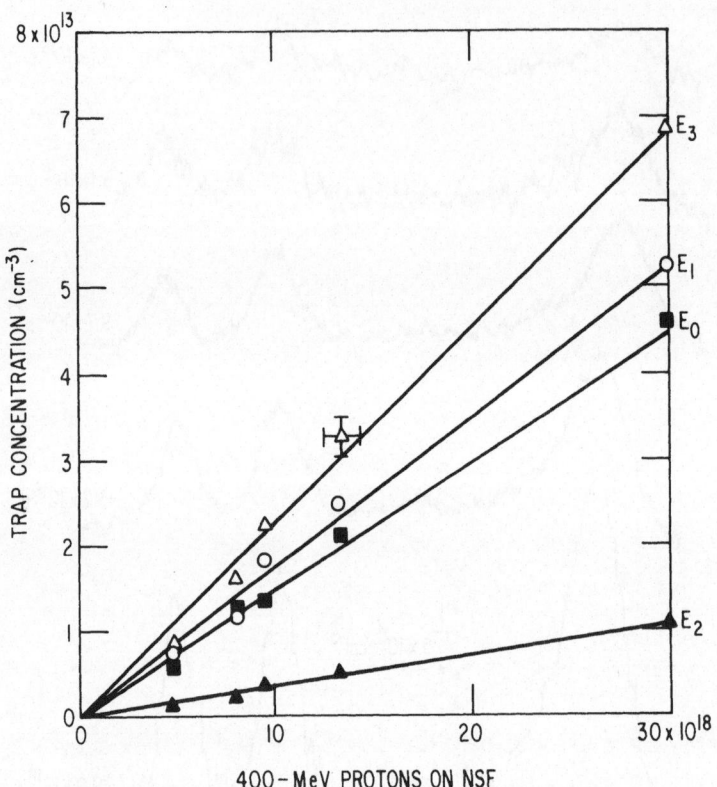

Fig. 11. Trap concentrations obtained from Fig. 10 vs. proton fluence on NSF target.

Figure 12 shows the results of 5 K neutron irradiation (REF target) and isochronal annealing on a different silicon sample. These results are similar to Kimerling's results for proton irradiation of Czochralski silicon[36] He associates the E_1 level with vacancy-oxygen complexes (A-centers) and the E_2 and E_3 levels with divancancies. The E_0 level has not been reported previously. However, this level has also been observed following room temperature reactor irradiations at the Univ. of Missouri. It should be noted from Fig. 12 that the annealing of the E_2 and E_3 levels are quite different. From Fig. 11 it is apparent that the production rates of these two levels are also different. It has also been reported at this conference that the profiles of these two levels as a

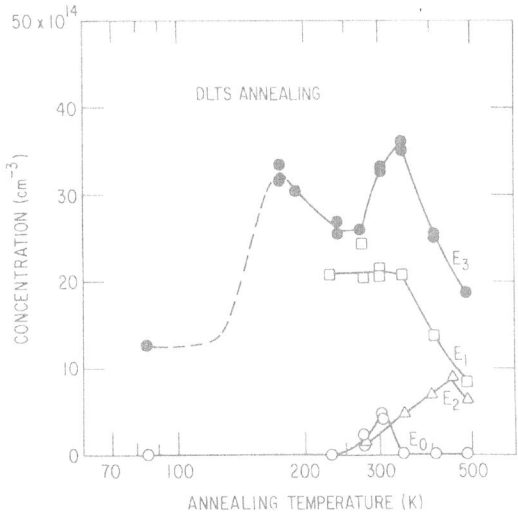

Fig. 12. Isochronal annealing (5 min.) of majority carrier traps in n-type silicon irradiated in REF at 5 K. Traps E_0, E_1, E_2, and E_3 are identical to the traps similarly labeled in Fig. 10.

function of distance from the junction also differ.[38] All of these facts suggest that one if not both of these levels have been misidentified or are associated with different defects in the case of neutron irradiation.

It should also be noted that the present experiments have not exhibited the low temperature DLTS peaks observed by Kimerling[36] following proton irradiation. One possible explanation is that these levels are associated with hydrogen-defect complexes which might be expected for the proton irradiation experiment. Clearly, additional experiments are needed to clarify these differences between low temperature neutron and proton irradiations.

SUMMARY AND CONCLUSIONS

These initial experiments have shown that pulsed-neutron sources can be a valuable tool for obtaining new information about degraded-fission neutron damage in semi-conductors. The conductivity of degenerate semiconductors has been observed to decrease exponentially with fluence at 5 K for all materials studied. The "available carrier model" presented here is more successful than the "cluster model" used previously for non-degenerate neutron damage at higher temperatures. Furthermore, the observed initial carrier removal rates agree reasonably well with calculations of the displacement rates.

A number of annealing stages have been identified in silicon and the II-VI compounds and can be associated with known defect migration stages. In silicon, very low temperature annealing (~22 K) has been observed for the first time. It also appears from these experiments that the Tellurium vacancy in CdTe might be mobile below room temperature. If true, this would be the only known vacancy in a II-VI semiconductor to possess this property. The annealing of neutron irradiated GaAs is quite different from the low temperature electron irradiation annealing results and clearly deserves additional study.

DLTS has been used to observe a small but measurable leakage flux at 24 keV between the two targets at the Argonne Intense Pulsed Neutron Source. Also, annealing experiments as well as defect production rates suggest that previous assignments of energy levels to the divacancy may be in error. Finally, a number of low temperature levels which are seen after proton irradiation, do not appear following neutron irradiation suggesting that these levels are associated with hydrogen-defect complexes.[20]

ACKNOWLEDGEMENTS

It is indeed appropriate to acknowledge the support of T. H. Blewitt of Argonne who not only pioneered the investigation of helium temperature neutron damage to metals, but was also responsible for the radiation effects facility at IPNS which provided us the opportunity to extend fission spectrum neutron damage in semiconductors to helium temperatures for the first time. We also wish to acknowledge J. M. Carpenter of Argonne, who has devoted so much of his effort in designing and building IPNS, for many useful discussions and for providing one of us (J.M.M.) housing while these experiments were performed. We also wish to express much gratitude to D. Hill of Monsanto and D. C. Reynolds of WPAFB for supplying, upon very short notice, the semiconductor samples used in these experiments.

REFERENCES

1.   R. R. Coltman, T. H. Blewitt and T. S. Noggle, Rev. Sci.
      Instrum. 28, 375 (1957).
2.   J. W. Cleland and J. H. Crawford, Jr., J. Appl. Phys. 29,
      149 (1958).
3.   H. J. Stein, Phys. Rev. 163, 801 (1967).
4.   H. J. Stein, J. Appl. Phys. 39, 5283 (1968).
5.   H. J. Stein, J. Appl. Phys. 40, 5300 (1969).
6.   R. C. Birtcher, T. H. Blewitt, M. A. Kirk, T. L. Scott, B.
      S. Brown and L. R. Greenwood, J. Nucl. Materials, in
      publication.
7.   L. R. Greenwood, J. Nucl. Materials, in publication.
8.   M. A. Kirk and L. R. Greenwood, Neutron Transmutation Doping
      in Semiconductors, ed. by J. M. Meese, Plenum Press, N.Y.
      (1979), p. 143.
9.   G. H. Kinchin and R. S. Pease, Rep. Prog. Phys. 18, 1
      (1955).
10.   B. R. Gossick, J. Appl. Phys. 30, 1214 (1959).
11.   J. H. Crawford, Jr., and J. W. Cleland, J. Appl. Phys. 30,
      1204 (1959).
12.   H. J. Juretschke, R. Landaner and J. A. Swanson, J. Appl.
      Phys. 27, 838 (1956).
13.   H. J. Stein and R. Gereth, J. Appl. Phys. 39, 2890 (1968).
14.   N. F. Mott, Rev. Mod. Sci. 40, 667 (1968).
15.   M. Caillot, Lattice Defects in Semiconductors 1974, ed. by
      A. Seegar, Inst. of Phys., London (1975), p. 280.
16.   J. M. Cleland and J. H. Crawford, J. Appl. Phys. 29, 149
      (1958).
17.   J. W. Corbett, Radiation Effects in Semiconductors 1976, ed.
      by N. B. Urli and J. W.Corbett, Inst. of Phys., London
      (1977), p. 1.
18.   H. J. Stein, Radiation Effects in Semiconductors, ed. by J.
      W. Corbett and G. D. Watkins, Gordon and Breach, New York
      (1971), p. 125.
19.   G. D. Watkins, Lattice Defects in Semiconductors 1974, ed.
      by A. Seeger, Inst. of Phys., London (1975), p. 1.
20.   L. C. Kimerling, P. Blood and W. M. Gibson, Defects and
      Radiation Effects in Semiconductors 1978, ed. by J. H.
      Albany, Inst. of Phys., London (1979), p. 273.
21.   This can be seen by an inspection of Table 18. Also see G.
      D. Watkins discussion of this subject in Defects and
      Radiation Effects in Semiconductors 1978, ed. by J. H.
      Albany, Inst. of Phys., London (1979), p. 16.
22.   G. D. Watkins, Radiation Effects in Semiconductors 1976, ed.
      by N. B. Urli and J. Corbett, Inst. of Phys., London (1977),
      p. 95.
23.   C. N. Elsby and J. M. Meese, IEEE Trans. Nuc. Sci. NS-21, 14
      (1974).

24.   F. L. Vook, Phys. Rev. $\underline{B3}$, 2022 (1971).

25.   M. Kitagawa and T. Yoshida, Appl. Phys. Lett. $\underline{18}$, 41 (1971).

26.   D. C. Look and J. M. Meese, Rad. Effects $\underline{22}$, 229 (1974).

27.   M. Caillot, Phys. Lett. $\underline{38A}$, 2 (1972).

28.   F. J. Bryant and E. Webster, Phys. Stat. Solidi $\underline{21}$, 315 (1967).

29.   F. J. Bryant, A. F. J. Cox and E. Webster, J. Phys. C $\underline{1}$, 1737 (1968).

30.   T. Taguchi and Y. Inuishi, J. Appl. Phys. $\underline{51}$, 4757 (1980).

31.   D. V. Lang, Radiation Effects in Semiconductors 1976, ed. by N. B. Urli and J.W.Corbett, Inst. of Phys. London (1977), p. 70.

32.   K. Thommen, Rad. Effects $\underline{2}$, 201 (1970).

33.   H. J. Stein, J. Appl. Phys. $\underline{40}$, 5300 (1969).

34.   A. H. Kalma, R. A. Berger, C. J. Fischer and B. A. Green, IEEE Trans. Nuc. Sci. $\underline{NS-19}$, 209 (1972).

35.   S. T. Picraux, Rad. Effects $\underline{17}$, 261 (1973).

36.   L. C. Kimerling, P. Blood and W. M. Gibson, Inst. of Phys. Conf. Ser. 46 (1979), p. 273.

37.   J. M. Meese, M. Chandrasekhar, D. L. Cowan, S. L. Chang, H. Yousif,
      H. R. Chandrasekhar, and P. McGrail, Neutron Transmitation Doped Si, ed. by T. Guldberg, Plenum, N.Y. (1981), p. 101.

38.   J. W. Farmer and J. C. Nugent, this conference.

FUTURE REACTOR CAPACITY FOR THE IRRADIATION OF SILICON

T.G.G. Smith

Harwell Atomic Energy Research Establishment
UK

ABSTRACT

The best estimates for future requirements for irradiation capacity are discussed and the factors which could influence them are enumerated.

The present world capacity of research reactors, with suitable facilities for the production of NTD silicon, is estimated and the variation of this capacity in the 10/15-year period is examined. The examination takes account of the ageing, safety regulations and economic influences.

The feasibility of constructing reactors dedicated to the production of NTD silicon is analysed in terms of technical and economic viability and the practicality of such a proposal is examined.

INTRODUCTION

The irradiation of large mono-crystal ingots of silicon with thermal neutrons in reactors, for the purpose of transmutation doping with phosphorus, has been proceeding, relatively troublefree, for over five years. One of the interesting aspects of the process was the rapid transition from a scientific concept, through the experimental phase to quantity production. This rapid transition was due, in part, to the relative simplicity of the process. The radiation work is repetitive, virtually no scientific or experimental input is necessary, and, once the basic parameters are established, all that is necessary is measurement, monitoring and control.

The irradiations are carried out in reactors of the research or experimental type because it is only in these reactors that neutron flux densities can be obtained which are sufficiently high to give short irradiation times and consequent short turn-round times. Additionally, it is only in reactors of this type that facilities exist, or can be installed, which permit the removal of the material, from the neutron flux, at the precise time when the desired level of phosphorus concentration has been achieved. However, non-scientific work of this nature is in stark contrast to that normally carried out in and around research reactors where, generally, each experiment is novel and requires scientific interpretation and innovation. Since the motivation to carry out the work is not scientific it must lie elsewhere and it is, of course, economic or financial.

In earlier years only a very few research reactors were operated on a commercial basis, but, in recent times, economic pressures have forced most reactor operators to examine their scientific work programmes and to compare them with operating costs. This examination has led to the closure of certain reactors as funding for scientific programmes declines in the face of world recession and, no doubt, will lead to the closure of others. It is fortunate that the advent of ND silicon coincided with this period of cost conciousness and provided the incentive for reactor operators to invest money in equipment and obtain income from the irradiation of silicon. However, as Bobby Stone pointed out in 1978[1] this marked the first time that a second party had become involved in the silicon production process. It must be recognised, therefore, that the process is completely reliant on thermal neutron densities and volumes, associated with the facilities which are only found in research reactors. It seems appropriate, therefore, to examine the total irradiation capacity which exists, or could be made available, to compare this with demand and to make a projection of capacity and demand for the next 15-20 years. However, before making this examination, it would be useful to examine the economics of the process from the reactor operator's viewpoint.

IRRADIATION ECONOMICS

First, let us examine the cost of operating an average research reactor:

If capital cost or depreciation of the plant is ignored then costs resolve into two main components:

(1) Operating costs, ie, salaries, wages, support services, equipment and consumables plus overheads

(2) Costs of nuclear fuel.

Typically, for a medium power reactor these are:

Operating costs	$5M/annum
Fuel costs	$1.5M/annum
TOTAL	$6.5M/annum

If it is now assumed that the price charged for irradiations is 6-7 cents/grm the rather alarming conclusion is reached that, for silicon alone to support the upkeep of a reactor of this type, 100 tonnes/annum must be irradiated. In fact, quantities a tenth of this amount are more realistic and so it becomes clear that silicon irradiation must always be carried out in conjunction with a large proportion of other work in a reactor and also, that a reactor dedicated to silicon irradiation is not a viable proposition.

FUTURE DEMAND AND SUPPLY

As was stated previously ND silicon relies on the availability of research reactors; an attempt will therefore be made to make an estimate of future demand and to relate this to irradiation capacity.

DEMAND

To make an estimate of future demand for ND silicon is extremely difficult, especially if a time-scale into the next decade is being considered. In 1980 H Hertzer made an estimate of future demand[2] and this is shown in Figure 1. In the two years since that estimate the indications are that it was slightly optimistic but it should be remembered that during those two years the world has experienced a trade recession of considerable magnitude. It should also be borne in mind that in the face of this continuing recession the demand for ND silicon has continued to rise whereas that for conventional material has been falling. Whilst opinions vary, there is a general agreement that the end of the recession could bring about considerable increases in demand. Figure 1 also shows the quantities which would be required, at various times in the future, for different rates of market growth, 3%-5% and 7% per annum. To predict the exact demand is impossible, but, with a slow economic growth, the demand in the 1990s could be 75-80 tonnes per year, whereas with higher world economic growth and the reasonable probability of innovation in applications, demand could be in the range 100-120 tonnes[3].

IRRADIATION CAPACITY

In attempting to assess the total reactor capacity presently available, and likely to be available up to the year 2000, a survey has been made of existing research reactors. The survey is based on

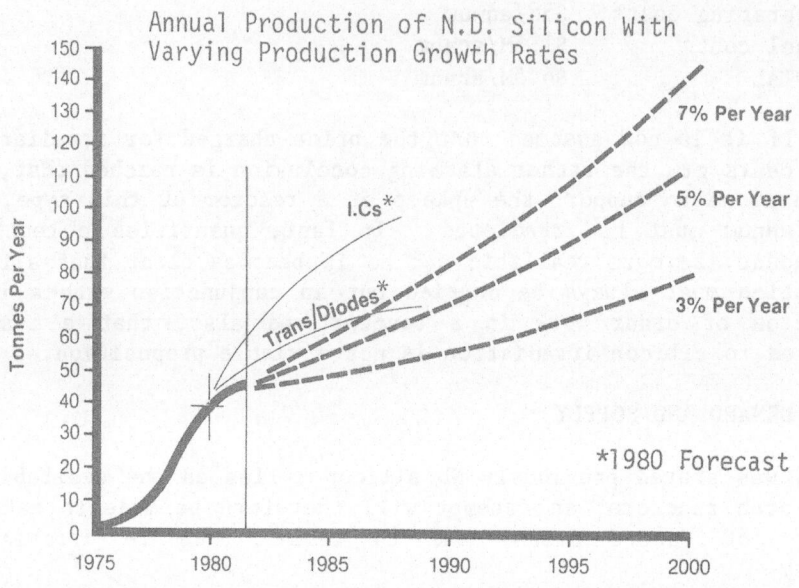

Fig. 1.   Predicted Production Growth Rates of NTD Silicon

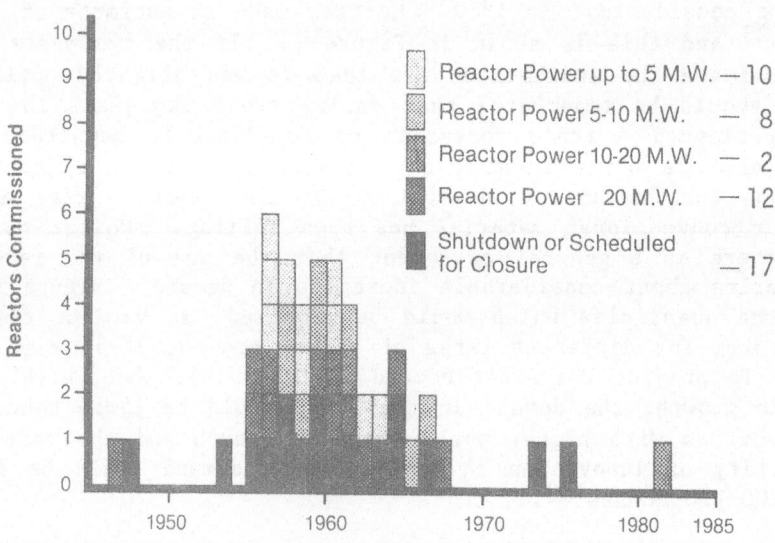

Fig. 2. World Reactors suitable for NTD Irradiations

the list of reactors published by the IAEA[4] but certain reactors have not been included for the following reasons:

Reactors with an operating power less than 3.5MW
Reactors dedicated to particular research tasks, ie, safety transient experiments etc
Reactors physically unsuited for the loading and unloading of silicon, eg, prototype power reactors
Reactors which, for political or policy reasons, are unlikely to become involved.

In Figure 2 the various reactors and their dates of commissioning are shown, together with those reactors which have been taken out of service, dismantled, or are scheduled for closure. The reactors are also broadly classified in groups according to power as this is an indication of available neutron flux.

It will be noted that almost all of the reactors date from the period around 1960 and only one reactor has been commissioned since 1968, the ORPHEE reactor commissioned in France in 1981. At the other extreme, the NRX reactor in Canada was commissioned in 1947 and is still in operation 35 years later.

At the present time there are few known plans to build new neutron sources and such as are being considered are mostly for "spallation" type machines, which do not appear to be suitable for ND silicon work.

By taking account of restraining parameters such as:

Neutron flux density
Physical layout constraints
Experimental programmes
Declared capacity[5]
Management policy
Known future plans
etc

it is possible to make an approximate estimate of the total capacity available for silicon irradiation work.

Figure 3 shows the estimated capacity and for convenience, this has been divided into three phases. Phase 1 shows the actual capacity which has existed since the work commenced and up to the present time. This capacity has closely followed demand and reflects the provisioning of special equipment and installation of facilities for handling silicon at the reactor stations. During this period capacity has always been slightly in advance of demand but if demand had increased then capacity could also have been increased. In Phase 2 the potential capacity up to the early 1990s

is shown.   In order to determine this on a world scale many
assumptions have had to be made.   It is considered unlikely that
every reactor which could be involved will, in the event, irradiate
silicon and the probability is that the bulk of the work will be
undertaken at the reactor centres already involved, augmented by
reactors not yet producing significant production quantities and
also by smaller quantities from the low power reactors.   Whilst the
total potential exists now in the form of reactors and neutrons, to
realise this potential would require the provision of a considerable
amount of equipment and therefore from the present day state could
only be achieved by about 1985.

In Phase 3, ie, the period beyond the early 1990s, a new
situation will arise.   By that time the reactors presently in use
will be approaching an average age of more than 30 years.   Figure 2
showed that of the reactors constructed in earlier years 35-40% have
already been shutdown and the rate of shutdowns will probably
increase as the age of the reactors increases.   There are also
influences other than age which are present and principal amongst
these is the ever-increasing standards demanded by safety and
regulatory bodies.   There are already cases of reactors being shut
down virtually without notice, and the GE reactor at Valecitos is a
case in point.   Recently the R1 reactor in Sweden suffered a similar
shutdown but is now operating again after modifications but on the
understanding that a major programme of refurbishing will be
undertaken in 1984.   The Swedish case demonstrates that ageing
reactors can be brought up to demanded standards, but, eventually
the cost of updating when compared against the present trend of
diminishing experimental programmes must result in reactor
closures.

COMPARISON OF DEMAND AND CAPACITY

In Figure 4, the demand predicted in Figure 1 and the capacity
predicted in Figure 3 have been superimposed and a direct
comparison is therefore possible.   Whilst the prediction of both
demand and capacity are subject to large uncertainty factors, they
do indicate that, at least until the mid-1990s, demand is unlikely
to exceed potential capacity.   Up to this time it seems likely that
actual capacity will be tailored to match demand and only if there
is a sudden and dramatic increase in demand will there be a
shortfall in capacity.   Beyond the mid-1990s there could be
inadequate capacity if the predictions for demand are realised.

OTHER NEUTRON SOURCES

The other major source of neutrons which will be in operation
up to the end of the century and beyond is nuclear power reactors.
It appears that the large majority of these will be of the
Pressurised Water type but, irrespective of type, they are designed

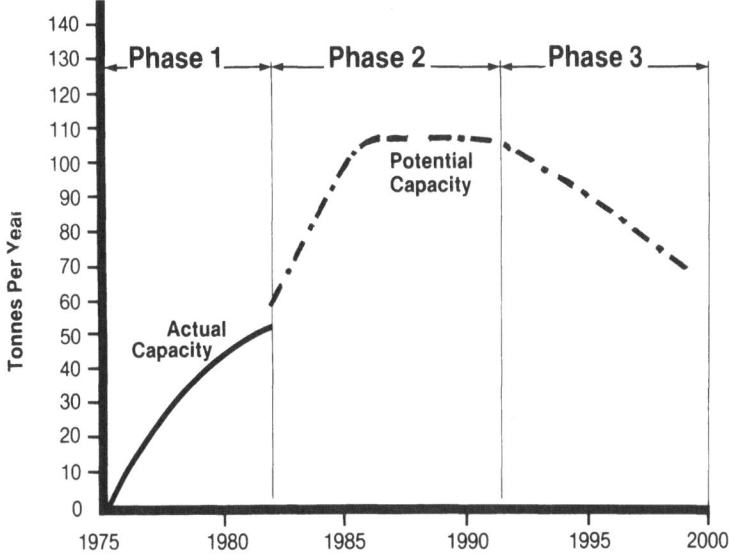

Fig. 3.   Predicted World Irradiation Capacity

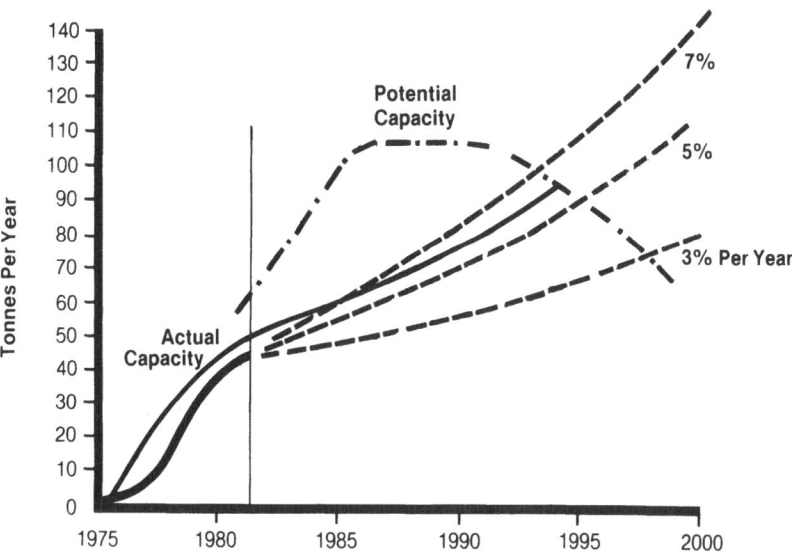

Fig. 4. Comparison of Irradiation Capacity and Demand to year 2000

exclusively for the production of power. The irradiation of silicon in these reactors would require changes and departures from a standardised design and it is highly unlikely that any operator could be persuaded to accept a non-standard design for the relatively small return which could be expected from silicon irradiations. Even a cursory examination of the economics demonstrates the impracticality of the proposal:

    Income from sale of electricity $250M/yr

If it is assumed that such a reactor could irradiate the whole world demand in the 1990s - say 100 tonnes - then the income would represent about 0.25% of the total income of the station. Such an involvement is unlikely to prove attractive when compared to the design costs involved and the non-standard type of operations required.

SUMMARY

On the basis of the assumptions made, and taking due account of the uncertainties in those assumptions, it emerges that for the next ten years there is more than adequate capacity available to cope with large increases in demand. Beyond this time the situation is less certain and declining capacity could fail to meet demand. However, it may be that the material produced by the process will be superseded by technological advances, or that conventional doping techniques will advance to produce dopant distributions comparable to those presently achieved by neutron transmutation doping.

REFERENCES

1.   BD Stone, "Large scale production of NTD Silicon", Jens Guldberg, Editor, Plenum Press, New York, p.20, 1981.
2.   H Herzer, "Neutron-Doped Silicon - A Market Review", Ibid, p.4.
3.   "Private communication, H Herzer-TGG Smith
4.   Research Reactors in Member States.   1980 Edition, International Atomic Energy Agency, Vienna.
5.   Paul Breant, Francois Cherrau, Jean-Pierre Genthon, "Development of the Irradiation Facilities for Silicon Neutron Doping in France", Jens Guldberg, Editor, Plenum Press, New York, p.300, 1981.

AN AUTOMATIC CONTROLLED, HEAVY WATER COOLED FACILITY FOR IRRADIATION

OF SILICON CRYSTALS IN THE DR 3 REACTOR AT RISØ NATIONAL LABORATORY,

DENMARK

K. Hansen, K. Stendal, K. Andresen and K. Heydorn

Risø National Laboratory
4000 Roskilde
Denmark

ABSTRACT

Neutron transmutation doping of silicon crystals with phosphorus has been carried out in the DR 3 reactor since 1975.

The irradiation capacity has been enlarged concurrently with the growing demand for neutron transmutation doped crystals.

This report describes a recently developed facility for the irradiation of 4 inches silicon crystals. The irradiation rig is placed in one of the vertical experimental tubes in the heavy water moderator of the reactor.

The rig is filled with heavy water, and the nuclear heat absorbed in the crystals during irradiation is removed by convection.

During the irradiation the crystals are placed in round aluminum containers. Transport of the containers from the irradiation zone to the top of the rig is made by heavy water flow. The irradiation cycle is controlled automatically by instrumentation.

1.   INTRODUCTION

Pure silicon mono-crystals doped with phosphorous atoms for adjustment of the electrical resistivity to a specified value are used in the electronic industry for production of thyristors, integrated circuits, etc.

In the conventional method of doping phosphorous atoms are added during the crystal growth and temperature nonuniformities and dopant segration produce significant variations of the resistivity across the crystal, whereas doping of silicon crystals with phosphorous atoms by transmutation gives a small variation in the resistivity over the crystal.

Nonradioactive phosphorous atoms are formed from the ^{30}Si isotope by the following process:

$$^{30}Si + n \rightarrow \ ^{31}Si \rightarrow \ ^{31}P + e^-$$

The content of ^{30}Si isotope in natural silicon is 3.09% and the half-life of decay of ^{31}Si is 2.6 hours.

At Risø National Laboratory in Denmark, doping of silicon crystals by means of thermal neutrons was started in 1974 for the Topsil Co., in order to prove the validity of the method and to provide test samples. The first irradiations were made in the thermal column of the DR 2 reactor (MTR-tank type) without special equipment.

The promising results from the test samples resulted in a demand for silicon crystals doped by neutron irradiation and a rig designed for irradiation of 3 inches silicon crystals in the DR 3 reactor was then manufactured during 1975.

Due to demand for further irradiation the facility was enlarged with four 3 inches rigs, manufactured during 1977. Finally in 1981 a facility for irradiation of 4 inches crystals was ready for operation.

This report describes the design and operation experience of the 4 inches facility.

2.   DESCRIPTION OF THE FACILITY

2.1 The Irradiation Position in the Reactor

The Danish DR 3 reactor is a 10 MW heavy water moderated and cooled research reactor. A vertical cross section of the reactor is shown in Figure 1. The 4 inches irradiation rig is placed in one of

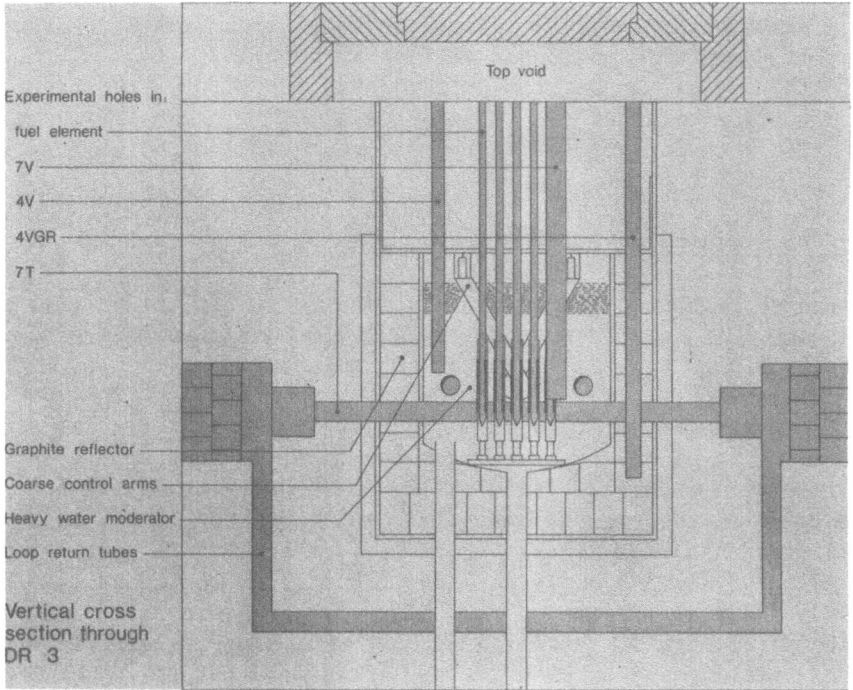

Fig. 1.   Irradiation Positions in the DR 3 Reactor

the vertical experimental holes in the heavy water tank. The thermal neutron flux, smoothed by the absorber screen, is about 2×10^{13} $n/cm^2 sec$. and the γ-heat in the irradiation zone is about 250 mW/g. Due to the γ-heat, it is necessary to cool the silicon crystals during irradiation. The temperature of the crystals during the ir- radiation is about $50^{\circ}C$.

2.2 Specifications for the Rig

The rig is designed for irradiation of silicon crystals of a diameter up to 107 mm and length up to 400 mm.

Variation in the phosphorous doping produced in the crystal is designed to be less than +/- 5% and this is routinely achieved in practice.

2.3 Methods Used for Obtaining Uniform Doping

In the irradiation zone of the 7V hole the variation of the in- cident thermal neutron flux is about 20% in axial direction and about 25% in radial direction.

In order to reduce the variation in the received neutron dose the following methods are used:

- An absorber screen of stainless steel with varying thickness is used to equalize the axial flux profile.
- The crystal is rotated at a constant speed of 2 rpm around the vertical axis of the irradiation hole.
  The rotation eliminates the influence of the radial flux gradient.

2.4 Crystal Container

The unirradiated crystals are loaded in a thin walled (0.5 mm) round container made from Al-2S. The inside diameter of the container is 110 mm and the length is 405 mm.

The container is strongly radioactive after the irradiation; but most of the activity comes from the ^{28}Al isotope with a short half life (2.3 minutes). After a 30 minute decay under the top shield plug of the rig the container can be removed from the rig by the shielded handling flask. The container and crystal are dried in the flask for about 2 hours and then transferred to the storage facility for further decay for 4-5 days. The remaining activity is now acceptably low for handling without shielding in connection with control, packing and shipment of the irradiated crystal.

2.5 Irradiation Rig

During irradiation the container is placed at the bottom of the rotating aluminum tube, which is supported by a ball bearing at the top and a graphite bearing at the middle.

The tube is rotated by a geared electric motor placed at the top of the rig. The rotating tube is surrounded by a rig thimble, which acts as containment for the heavy water in the rig. The absorber screen is placed on the outside of the rig thimble. The rig has a top shield made from stainless steel with channels for the heavy water flow used to transport the container up from the rig. A lead filled stainless steel plug is placed in the top of the rotating tube. During loading operations the plug is removed and stored in inside the handling flask.

Figure 2 shows the rig placed in the reactor.

2.6 Heavy Water Circuit

The heavy water circuit is used for transport and indication of containers in the rig.

Fig. 2.   The Irradiation Rig Placed in the Reactor

The main components of the circuit are:

- The irradiation rig
- The levelling tank
- The pump unit

The pump unit consists of pumps, valves, ion exchanger, filter, drain tank and transducers for control of flow, pressure and conductivity.

Transport of the Container. When the irradiation of the crystal is finished one of the main pumps is started by a signal from the flux integrator in the instrumentation. The heavy water flow circulating up through the rotating tube and forces the container up to the top of the rig. The water flow is about 75 litres/minute and the gap between the container and the tube is 1.5 mm.

The container is stopped by the shield plug at the top of the rig. The transport time from bottom to top is 1-1.5 minutes. The container is then "cooling" for about 30 minutes at the top of the rig. The main pump is first stopped when the irradiated container is removed to the handling flask and an unirradiated container is placed in the top of the rig.

When the main pump is stopped, the container will sink down in the rotating tube and reach the bottom after 0.5-3 minutes depending on the crystal weight.

Indication for a Container in Irradiation Position. A nozzle is placed in the bottom of the rig. The bottom end plate of the container will close the nozzle, when the container is in irradiation position.

The indication pump (diaphragm type) forces a water flow (about 1.5 litres/minute) through the nozzle. The delivery pressure of the pump indicates when a container is in irradiation position. The indication pump is running continuously and is also acting as a cleaning pump by sending the water through a filter and an ion exchanger.

The heavy water from the rig and the circuit can be drained to the drain tank by means of the indication pump.

The pump unit is shown in Figure 3.

2.7 Handling Flask

The shielded handling flask (see Figure 4) is used for loading opations in the rig and in the storage facility and for transfer operations between the rig and the storage facility.

The flask is carried by a small wagon driven by a geared electric motor. The shield is formed as a cylindrical jacket with a thickness of 12 cm lead. A drum with a loading hole and positions for the rig plug, an irradiated container and an unirradiated container is placed inside the jacket.

The plug and the containers are lifted by a pneumatically activated grab during loading operations. The grab is supplied with compressed air through a reinforced rubber hose, which also is used as rope when the grab is lifted by a winch placed on the top of the flask.

By means of an indication valve it is possible to control when the plug or a container are correctly placed in the grab. The order of the single operations are controlled by a special guide disc placed on the top of the flask.

Fig. 3.   The Pump Unit

Fig. 4.   The Shielded Handling Flask

The wet containers are dried in the flask for about 2 hours by means of heated air circulated through the container. The evaporated heavy water is recovered in a condenser. All inside surfaces in the flask are stainless steel in order to avoid rust contamination on the containers.

2.8 Storage Facility

After irradiation it is necessary to place the containers in a shielded storage facility 3-5 days before they can be handled without shielding. The duration is determined by the ^{72}Ga impurity in the aluminum of the containers.

The storage facility consists of 23 stainless steel tubes arranged in five rows and shielded by concrete containing lead shot and iron shot.

The top of the storage facility is in level with the reactor top (see Figure 5). Loading operations are made by means of the handling flask, which can be moved on the reactor top between the rig and the storage positions.

Fig. 5.   The Storage Facility

2.9 Instrumentation

Automatic Dose and Transport Control. The thermal neutron flux is not constant during the time required for the irradiation. It is therefore necessary to integrate the thermal flux over the irradiation time in order to obtain acceptable accuracy of the received neutron dose.

The thermal neutron flux is measured by three self-powered neutron detectors of the vanadium type placed in guide tubes in the irradiation zone of the rig.

The signal from one of the neutron detectors is connected to an analog to digital converter, which integrates by means of a frequency proportional to the signal from the neutron detector. Before irradiation the integrator is set at a predetermined value.

When the container reaches the bottom of the rig a rise in the indication pressure starts the integrator and the rig rotation. The integrator counts downwards during the irradiation and at zero count the integrator starts the main pump and the container is lifted away from the irradiation zone by the heavy water flow.

Rotation Control. Control of the rig rotation is made by means of an opto-electrical detector activated by slits in a small aluminum disc.

Warnings. Warnings for abnormal conditions in the rig system during irradiation are announced by sound and by lamps on a display in the instrumentation rack.

Warnings during irradiation:

- Stop of rig rotation
- High and low level in the levelling tank
- Diaphragm fault in the indication pumps
- Low main flow
- Low pressure and flow in the indication system

Warnings at the end of irradiation:

- End of irradiation
- Low main flow
- Transport error (container not lifted from bottom of rig)

The warnings are connected to the "experiment warning guard" in the control room of the reactor.

The instrumentation panel is shown in Figure 6.

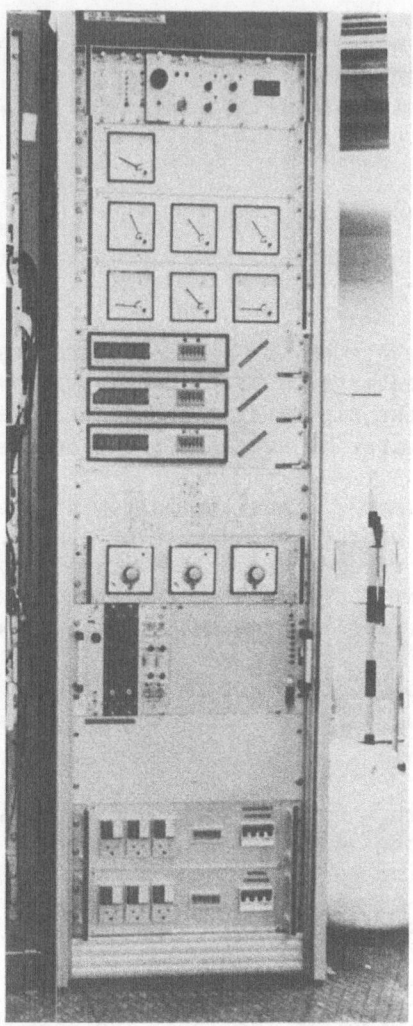

Fig. 6.   The Instrumentation

3.   OPERATION EXPERIENCE

The facility was ready for operation in August 1981.

The planned irradiations have been performed without significant problems.

3.1 Experience with Use of Heavy Water in the System

Heavy water is used in the system due to its low neutron absorption property. Measurements have shown that use of light water

would reduce the neutron flux level more than 50% and give us difficulties in equalizing the axial flux profile.

The disadvantages of heavy water are high cost and tritium
build-up during irradiation.

Losses of heavy water. Due to the high cost of heavy water it
is important to minimize losses during operation.

Evaporation losses are prevented by having a tight system.
Only during loading operations the rig is open to the reactor top.

The irradiated containers still contain a small amount of heavy
water when they are removed from the rig. The containers are dried
in the handling flask as mentioned in Section 2.7 and the heavy
water is recovered in the condenser, but it is degraded to about 80%
by light water from the atmosphere.

Tritium build-up. The system was filled with "fresh" tritium
free heavy water at the start in August 1981. A slow build-up of tri
tium takes place due to neutron irradiation.

Calculations have shown that the heavy water can be used for at
least 2 years before the tritium concentration exceeds the tolerated
level and exchange of the heavy water in the system is required.

3.2 Handling of Crystals

The crystals are protected by the aluminum containers during
transport and irradiation. All operations in connection with loading and unloading of crystals in the containers are made in a
special equipped room in the isotope laboratory, shown in Figure 7.

After opening of the irradiated containers the crystals are
rinsed in demineralised water and, if required, ultrasonicly cleaned; no crystals have as yet been strongly contaminated.

3.3 Accuracy of the Neutron Dose

In order to obtain a quality control on the neutron dose two
cobalt monitors are placed in each container, one at the bottom and
one at the top.

Results have shown that the flux integrators are working pre-
cisely and are maintaining the neutron doping within the specified
values.

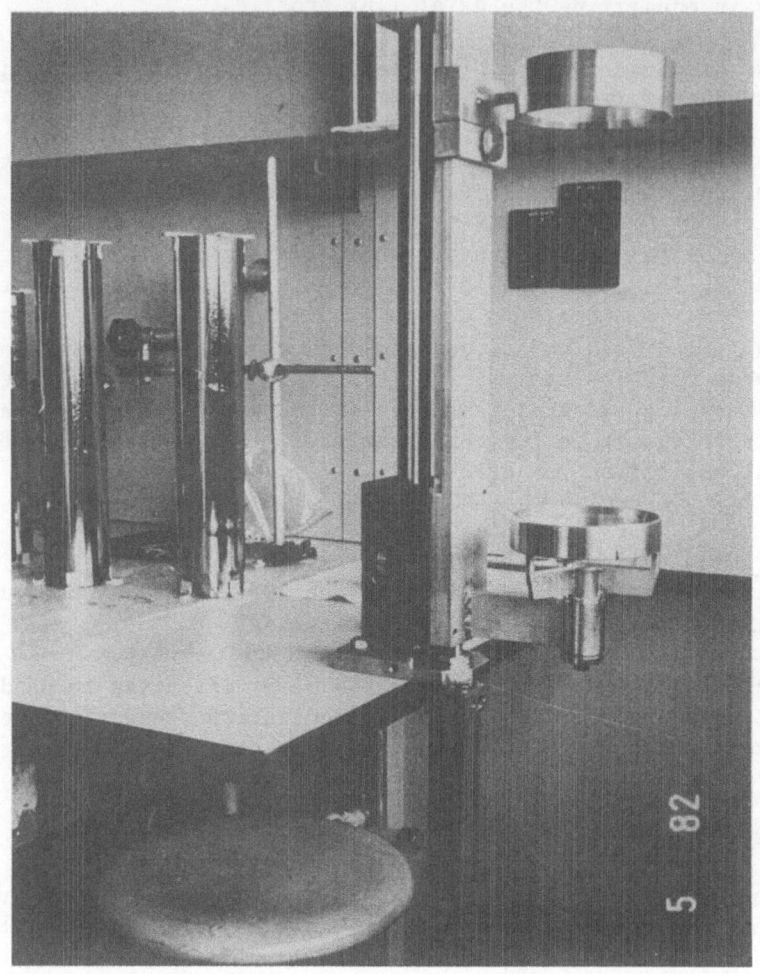

Fig. 7.   The Packing Room in the Isotope Laboratory

4.   CONCLUSION

The described facility has been in operation since August 1981, and the planned irradiations have been performed without significant problems.

The relative high thermal neutronflux in the irradiation position and the high utilization (≈ 83%) of the DR 3 reactor give the facility a considerable irradiation capacity.

IRRADIATION OF SINGLE SILICON CRYSTALS WITH DIAMETERS IN THE 3-TO 5-INCH RANGE IN FRENCH REACTORS

Paul Bréant

Commissariat à l'Energie Atomique

Paris, France

ABSTRACT

Neutron-doped silicon is widely produced in France, mainly in swimming-pool research reactors. New irradiation facilities adapted to 4- to 5-inch diameter ingots have been constructed at several locations. These facilities use neutronic absorption screens to produce a uniform axial distribution of thermal neutron flux. This feature, plus the systematic calibration of the irradiation equipment have contributed to the attainment of the target resistivities to within a mean deviation of -4 to +1.7 percent.

INTRODUCTION

The various methods used for the neutron transmutation of silicon 30 into phosphorus 31 have been studied systematically. These methods are well documented and easily controllable and have led to the utilization of specific facilities in many reactors in the world.

In the various conferences held on this subject in Philadelphia (1977), the University of Missouri (1978), and Copenhagen (1980), much attention was given to the different methods employed by reactor operators to satisfy both the quantitative and qualitative requirements of the silicon manufacturers.

France, which has a wide range of reactors, and, more importantly, research reactors, at her disposal, has made significant contributions towards developing and improving irradiation procedures. Attention was drawn at the Copenhagen conference to the wide range of characteristics that French irradiation facilities have to offer: pool reactors with and without heavy water reflectors, an extremely wide range of thermal flux/fast flux ratios, a large number of irradiation possibilities in many reactors in several research centers, continuous production and experience already acquired with previous customers.

In 1980, most of the available irradiation devices had the following geometrical limits: maximum length of between 25 and 50 cm, diameter generally limity to 3 inches. These limits corresponded to customers requirements.

During 1982 these requirements changed somewhat and diameters between about 80 mm and 104 or even 110 mm (4 inches) were sought. A significant amount of materials of even greater diameter (up to 5 inches) is now appearing on the market.

The purpose of the present paper is to summarize recent work conducted in France that has led to immediate availability of significant irradiation capacities for diameters up to 130 mm.

It is worth noting that the almost exclusive use of pool type reactors in France favored the satisfying of increased diameters requirements. In fact, the only restriction encountered in this field is the time normally required to fabricate and calibrate the corresponding irradiation equipment.

This paper will discuss the level of our current capacity for irradiations in 3- to 5-inch diameter devices and will demonstrate the guarantees that can be assured with our reactors based on:

- measurements made by our reactor physics teams,
- irradiation results from technical reports established by our customers.

I. DEVICES FOR IRRADIATING SILICON WITH DIAMETERS IN
   THE 3-TO 5-INCH RANGE:

Neutron-doped silicon is produced in research
reactors in France located in two nuclear research centers, one at Saclay (near to Paris) and the other at
Grenoble. Facilities capable of irradiating large diameter single crystals of silicon exist in these reactors. It is worth noting that all the irradiation facilities have the following characteristics in common:

- an irradiation length of between 350 and 500 mm,
- fluence control by an array of self-powered detectors
  distributed along the lenght and across the diameter
  of the different samples,
- sample cooling in such a way as to ensure that the
  irradiation temperature is less than 100°C,
- an automatic control of the incident fluence by a
  computer system associated with the self-powered
  detectors and capable of correcting local flux variations due to the reactor so that the absolute fluence
  value sought can be obtained with a high degree of
  accuracy. The removal of the sample from the flux
  after irradiation is controlled automatically,
- an axial flux distribution corrected by special calibrated neutron screens. The relative positions of
  the samples and screens can be axially adjusted in
  order to take changes occurring in the reactor core
  (in which the irradiation is effected) into account,
- the assurance of a radial homogeneity of the fluence
  by rotating the sample,
- a sufficiently high instantaneous flux level for relatively short irradiation times to be used. The
  deactivating time is therefore limited to a minimum
  duration determined by the fluence required and
  the activity of the phosphorus 32.

This series of characteristics enables customer
requirements to be satisfied by assuring a ± 10% minimum guarantee for the fluence requested. In pratice,
the results obtained enable a much better figure to be
achieved (of the order of ± 4%).

The principal characteristics of French reactors
are summarized in table I. This table shows the level
of our current capacity and it should be noted that due
to the nature of our reactors, this capacity is not
volume limited.

TABLE 1

FRENCH SILICON IRRADITATION FACILITIES

$$\frac{\text{thermal flux}}{\text{fast flux}} \sim 10$$

Reactor : name	MELUSINE	OSIRIS	SELOE
type		swimming pool H_2O →	
location	Grenoble	Paris(Saclay)	Grenoble
Characteristics of the irradiation facilities :			
. maximum diameter of the Si ingot	4"	5"	2"
. possible extension of the ingot diameter	yes	yes	yes
. environment	H_2O	H_2O	H_2O
. thermal flux ϕth 10^{13} n cm^{-2} s^{-1}	1 to 5	5 to 10	5 to 10
. ratio ϕth/ϕr (ϕr rapid flux E > 1 MeV)	~ 10	~ 10	~ 10
. pratical and present capacity (tons/year for 5 x 10^{17} n cm^{-2})	9	9	(1)
. possible extension of capacity (delay < 6 months) (tons/year)	15	15	

$$400 < \frac{\text{thermal flux}}{\text{fast flux}} < 10\ 000$$

Reactor : name	ORPHEE	MELUSINE
type	D_2O H_2O cooled	swimming-pool D_2O tank
location	Paris (Saclay)	Grenoble
	to start in 1983	
Characteristics of the irradiation facilities :		
. maximum diameter of the Si ingot	4.5"	3"
. possible extension of the ingot diameter	no	no
. environment	He or CO_2 or H_2O	H_2O
. thermal flux ϕth 10^{13} n cm^{-2} s^{-1}	3 (to be confirmed)	0.3
. ration ϕth/ϕr (ϕr rapid flux $E > 1$ MeV)	10,000	600
. practical	(~ 10) to be specified	0.4
. possible extension of capacity (Delay < 6 months)	no	no

N.B.:

1. It is specified that all existing facilities are instrumented
   with ingot rotation, flux monitoring, automatic time integra-

Continued

TABLE 1. Continued

tion and are associated with handling, storage, cleaning, decontamination and activity control equipment. Concerning the irradiation capacities, it is pointed out that they are realistic and present capacities for presently running devices and according to the way they are used on an average; that is to say that they take into account the real average dimensions of the ingot supplied and not the maximum possible loading with the maximum diameters. Further extensions of capacity are possible if need be by the installation of supplementary standard irradiation modules in the pools.

2. Fast flux E > 1 MeV

II. NEUTRONIC STUDIES OF 3-TO 5-INCH IRRADIATION FACILITIES

A systematic program of development has enabled large diameter irradiation facilities to be developed.

This results has been achieved by interposing an absorption neutron screen between the sample and the neutron source to improve the uniformity of the axial thermal-flux distribution. This screen may be made of either stainless steel or nickel. The final dimensioning of the neutron screen is determined from dosimetry measurements:

- on a model of the experimental reactor. This is case for OSIRIS and SILOE, which each have a critical model,
- on the irradiation device itself in an experimental reactor.

Several parameters are controlled :

- Effect of the Vertical Positioning of the Screen

This first measurement is made using a small fission chamber located inside an aluminum block which replaces the silicon.

Aluminum and silicon have similar absorption and scattering cross sections.

Units: $10^3 \, cm^{-1}$	Silicon	Aluminum
Absorption cross-section	8	15
Scattering cross-section	84	89

Measurements are made when the reactor control rods are in a position corresponding to the xenon equilibrium in the core and when an equilibrium exits between the burnup of the soluble poison inside the fuel elements (boron) and the uranium 235 burnup such that control rods remain in an almost fixed position over longs periods of time. The position of the screen is then ajusted so that the axial distribution as flat as possible. In general, this corresponds to a position below the median plane of the core.

This adjustment must be accurate. Figure 1 shows that a displacement of only 10 mm in the position of the screen leads to a significant variation in the axial distribution of the thermal flux.

- Effect to Size of the Reactor Control Rods and Possible Modifications in the Axial Flux Distribution at the Corresponding Irradiation Position

Figure 2 demonstrates the effect of the position of the reactor control rods when the screen is held in a fixed position.

It is seen that it is important to have a movable screen which can correct any modifications in the axial flux distribution (and in particular those occuring at the end of the cycle). Such modifications may be due for example to a displacement of the control rods as a result of burnup of the fuel's soluble poison.

- Effect of the Filling up of the Irradiation Device: Diameter and Lenght of Samples

Calibrations are performed using a stack of polycrystalline silicon disks separated by thin spacers provided with gold integrators.

It is important that:

a) the sample is surrounded by the least possible volume of water. The scattering cross section of light water is in fact 40 times larger than that of silicon. It is also important that the ratios of the respective volumes occupied by the silicon or aluminum and the water remain constant. In order to do this, each of the silicon sample is housed in an aluminium tube, which has the same diameter independent of the diameter of the silicon sample.

b) a spacer is employed at the top and bottom ends of the sample. This spacer may be made of silicon or aluminum; its purpose is to avoid local pertubations in the flux.

A recapitulation of three principal dosimetry methods employed in France has shown that it is possible to use a thin neutron screen capable of accurately smoothing the thermal flux distribution at the place where large diameter silicon samples are irradiated (along the periphery of the pool reactor's core).

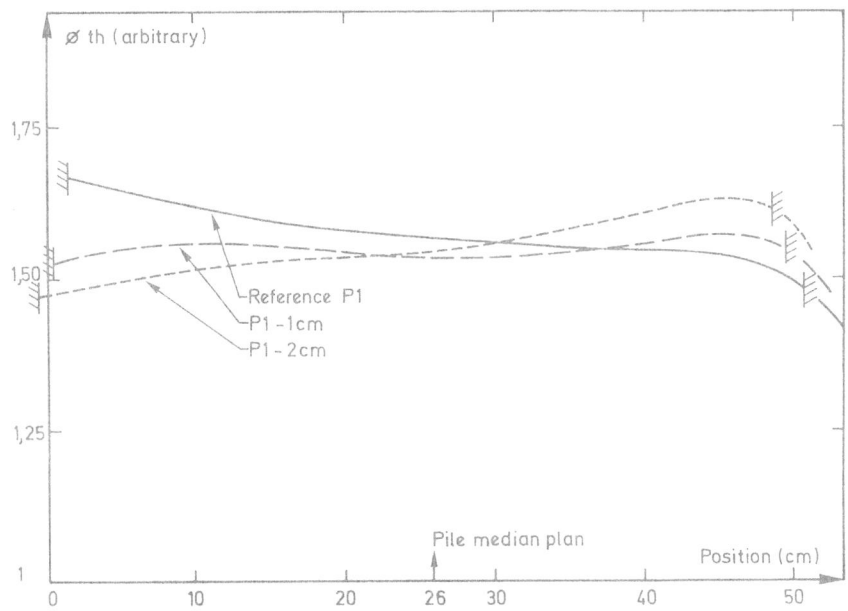

FIG. 1
NEUTRON SCREEN-POSITION INFLUENCE

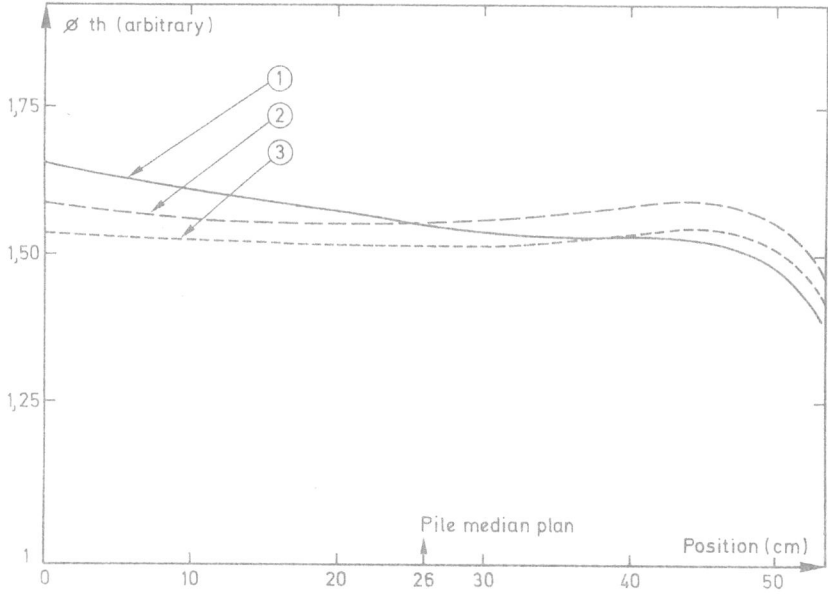

1   4,5 control rods up   FIG. 2
2   5 control rods up   CONTROL RODS POSITION INFLUENCE (OSIRIS)
3   5,5 control rods up   Neutronic Screen P1-1cm

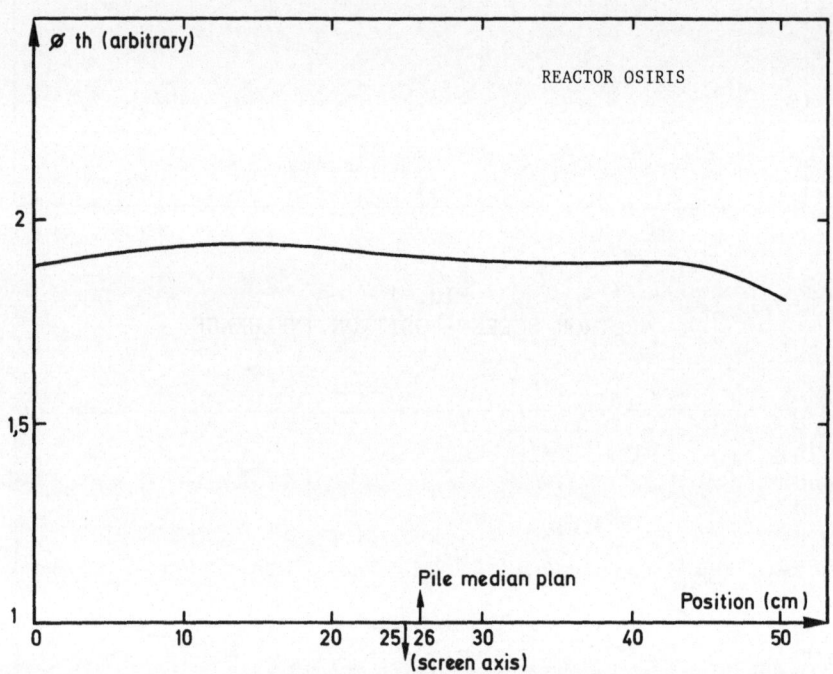

FIG. 3
AXIAL DISTRIBUTION (al ϕ 6 inches)

Figure 3 shows the actual results obtained at an irradiation position in the OSIRIS reactor at SACLAY. It can be seen that the irradiation distribution is constant to within 4% over more than 400 mm even though the source (reactor core) is located at a height of 600 mm.

III. TEMPERATURE CONDITIONS

It is obvious that the use of pools reactor greatly facilitates the cooling of the silicon samples.

The samples are directly immersed in the pool. Residual pollution problems have never occured. In any event, the samples are carefully rinsed after irradiation.

The maximum irradiation temperature is of the order of 70°C for a maximum gradient of 13°C in the silicon sample.

This level of performance is obtained with a maximum coolant flow rate of 1.25 m³/h and a mean water temperature of 35°C. The total γ power generated in the silicon sample is of the order of 7.5.KW.

IV. CONCLUSIONS

Over a period of several years, several tens of tons of silicons have been irradiated in French reactors. Significant quantities of large diameter samples have been irradiated since 1981.

Excellent results have been obtained, due in particular to the systematic calibration of the irradiation equipment used and also to the contributions that our customers have made to these studies. It is essential that a permanent two-way exchange of information is established so that any deviation which might occur can be corrected. The quantities of large diameter samples handled are now sufficiently significant for statistical analysis. The mean deviation about the absolute target resistivity specified by the customer lies between - 4% and + 1.7%. The dispersion recorded (standard deviation) is at the most 3.

References :

1. Morin, C., 1980, Test screen for silicon irradiation, note E.828 in French.

2. Bréant, P., Cherruau, F., Genthon, J.P., 1980, Development of the irradiation facilities for silicon neutron doping in France, 3rd International Conference on NTD of Silicon.

3. Maillot, R., 1982, Technical information sheet 616 (in French).

THE DEVELOPMENT OF NTD TECHNOLOGY IN THE INSTITUTE OF ATOMIC

ENERGY

Cao Yizheng and Gao Jijin

Reactor Engineering Division
Institute of Atomic Energy
Beijing, People's Republic of China

INTRODUCTION

The plan for developing NTD technology in the Institute of Atomic Energy began in 1979. Experiments on the irradiation conditions, irradiation facilities, and annealing after irradiation have been carried out at the Swimming Pool Reactor (SPR) at the Institute. The initial technology developed was successfully used for production of NTD Si[1]. Subsequently, the SPR was modified to install some additional irradiation facilities, and the commercial irradiation of NTD silicon ingots was initiated in 1980.

The capability for irradiating silicon was also included during the reconstruction of the Heavy Water Research Reactor (HWRR) at the Institute. In order to irradiate large diameter silicon ingots, four vertical tubes each with a diameter of 120 mm were installed in the heavy water reflector. Two sets of irradiation facilities along with transportation and cleaning equipment were in use in 1981.

IRRADIATION FACILITIES

The Swimming Pool Reactor

The cross section of the SPR core is shown in figure 1. The maximum thermal power of the reactor is 3.5 MW, and the maximum thermal neutron flux 3.5×10^{13} n/cm^2·s. Ten vertical irradiation tubes are available for irradiating silicon ingots, of which nine tubes with inner diameter of 65 mm are in the reflector, and the remaining tube with an inner diameter of 105 mm is adjacent to the

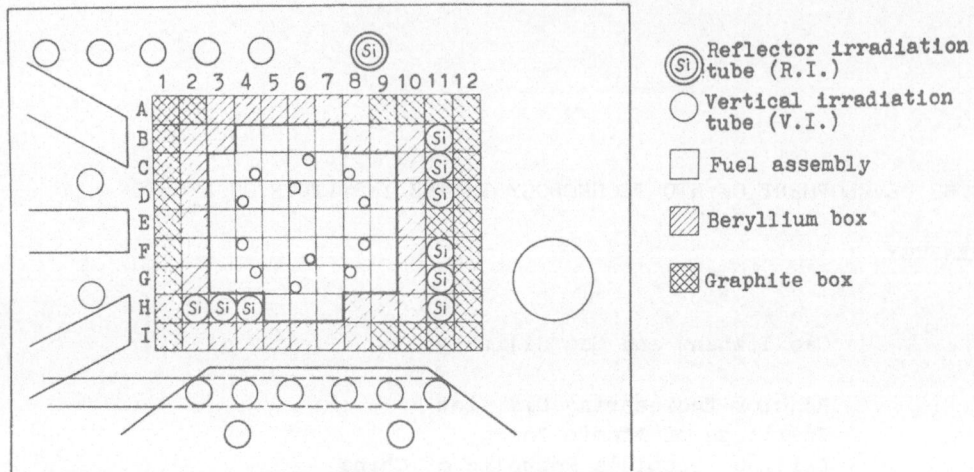

Fig. 1. Cross section of the Swimming-Pool Reactor (SPR).

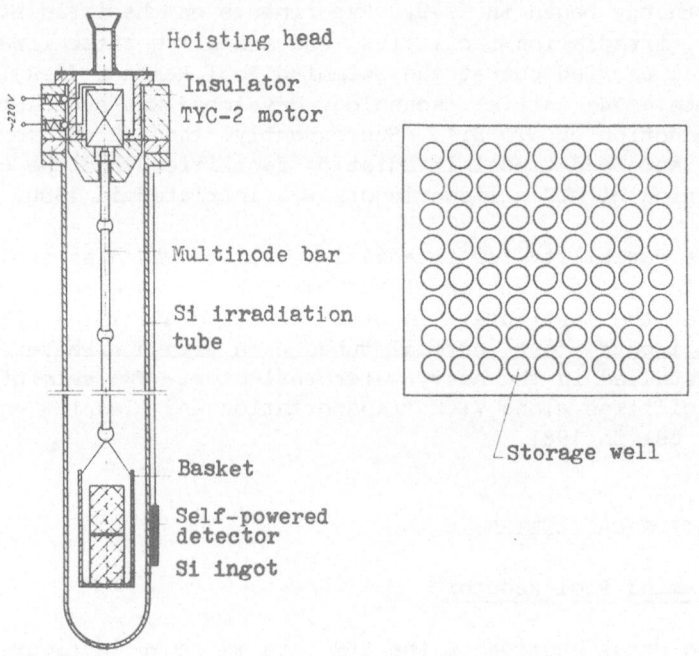

Fig. 2. Schematic diagram of the SPR irradiation facility and its storage well.

core. All these tubes are wet tubes filled with pool water. The irradiation facilities in this reactor are straightforward and shown in figure 2. The characteristics of these ten irradiation tubes are listed in table 1.

The storage well is used for temporary storage of the irradiation basket containing the silicon ingots. To ensure the axial uniformity, once the irradiation fluence reaches one-half of the total predetermined dose, the silicon ingots are turned upside down or exchanged in position with respect to each other by means of a manipulator located at one side of the reactor hall as shown in figure 3.

The Heavy-Water Research Reactor

The cross section of the HWRR core is shown in figure 4. The compact core has a central heavy-water cavity and an outer heavy-water reflector, each serving as a neutron trap. The 15 MW thermal power of the HWRR corresponds to a thermal flux peak of 2.5 × 10^{14} n/cm^2·s in the central cavity and a smaller peak of thermal flux in the outer reflector. Thirty-three vertical channels are arranged in the central cavity of the core and the outer reflector and can be used for production of radioisotopes, testing fuel elements, neutron activation analysis, and irradiation of NTD silicon ingots[2].

Two irradiation facilities for NTD of large silicon ingots have already been installed in the vertical channels located in the outer heavy-water reflector. One is a two-inch facility in channel No. 29 and the other is a three-inch facility in channel No. 24. Furthermore, a four-inch facility in channel No. 20 will

Table 1. Characteristics of the SPR Irradiation Facilities

Facility Name	V.I.	R.I.
Location	Water pool	Reflector
Number of tubes available	1	9
Thermal flux ϕ_{th}(n/cm^2·s)	2.4 × 10^{12}	(1-3.3) × 10^{13}
Ratio ϕ_{th}/ϕ_f(ϕ_f-fast flux E>1 MeV	19	10-20
Irradiation environment	H_2O	H_2O
Irradiation temperature of Si ingot (°C)	40	40
Maximum diameter of Si ingot (mm)	95	55
Maximum length of Si ingot (mm)	150	150

Fig. 3. Manipulator for exchanging position of ingots.

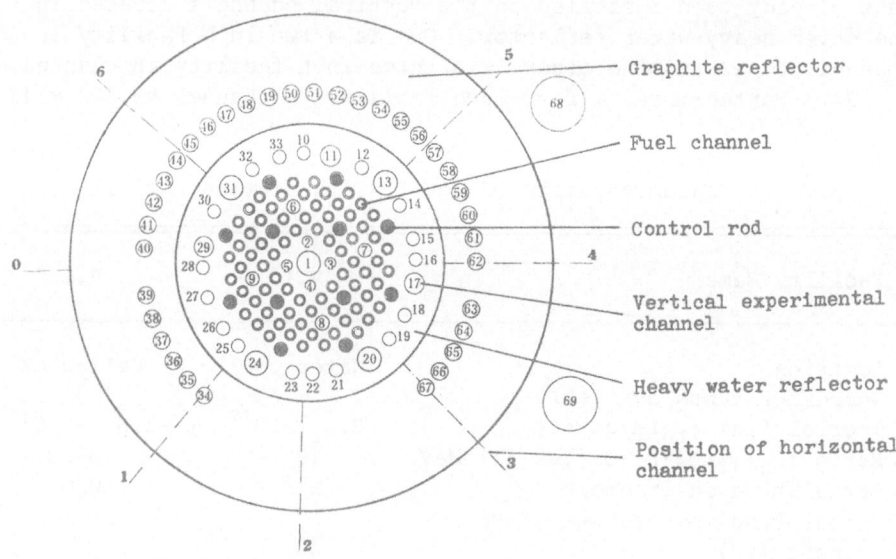

Fig. 4. Cross section of the Heavy-Water Research Reactor (HWRR).

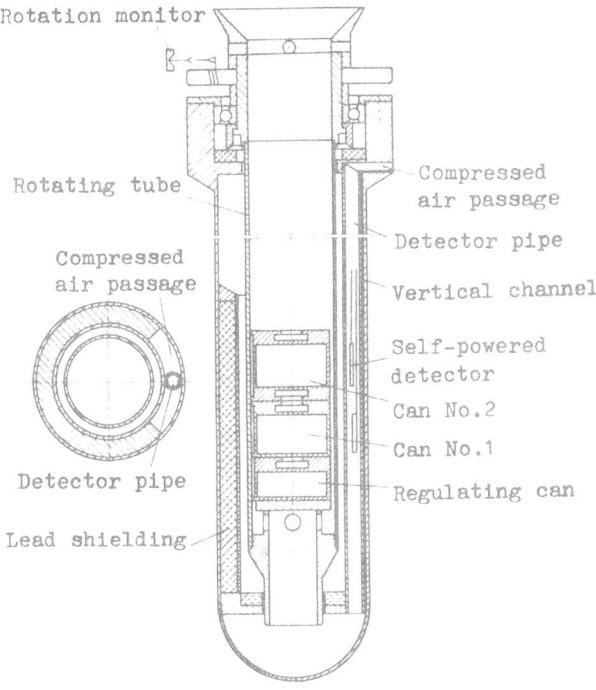

Fig. 5. Schematic diagram of irradiation facilities in the HWRR.

be installed in 1983. The schematic diagram of the irradiation
facilities is shown in figure 5.

All these irradiation facilities are isolated from heavy-
water. The air gap surrounding the irradiation cans makes it very
convenient to take cans out of or to put them into the facility.
The original remote and automatic handling device for isotope
targets was modified so it could be used for handling cans of NTD
silicon.

The high thermal neutron flux in the channel is accompanied
by intense gamma rays. In order to ensure the irradiation temperature of the silicon ingots less than 180°C, lead shielding with a
thickness of 1 cm is placed at the side adjacent to the core to
reduce the gamma heating in silicon ingots. A forced circulating
system of compressed air is used for cooling silicon ingots, the
flow of compressed air being about $1M^3$/min. The flowing air
passes through small holes on the side wall of the cans and di-

Table 2. Characteristics of the Irradiation Facilities
in the HWRR

Facility Name	2-inch	3-inch	4-inch
Location (see figure 4)	#29 Vertical Channel	#24 Vertical Channel	#20 Vertical Channel
Thermal flux ϕ_{th}(n/cm^2·s)	$(4-9) \times 10^{13}$	$(4-9) \times 10^{13}$	$(4-9) \times 10^{13}$
Ratio ϕ_{th}/ϕ_f(ϕ_f-fast flux E>1 MeV)	>50	>50	>50
Irradiation environment	air	air	air
Maximum irradiation temperature (°C)	<150	<180	<180
Maximum diameter of Si ingot (mm)	52	82	100
Maximum length of Si ingot (mm)	280	250	200

rectly blows on the surface of the silicon ingot to increase the
cooling efficiency.

As shown in figure 5, there are two cans in each facility and
two rhodium self-powered detectors used for real-time control of
the irradiation fluence of each can. Proper position of the irradiation cans in the facility is ensured by adjusting the height of
a regulating can consisting of an aluminum can filled with graphite.

The irradiated cans are stored in the cooling thimbles which
are located in the storage pool of the depleted fuel elements.
Then by a special transportation device the cans are transported
into a shielding box for disassembling. The main parameters of
irradiation facilities are listed in table 2.

IRRADIATION TECHNIQUE AND MAJOR SPECIFICATION

Control of Doping Accuracy

Determination of Doping Factor. The doping factor was determined
by using the Westcott cross section[3]. The concentration of dopant
phosphorus C_p, generated by neutron capture, is given by the
relationship:

$$C_p = N_o \hat{\sigma} \, nv_o t \qquad\qquad (1)$$

where N_o is the density of atoms of ^{30}Si per cm^3, nv_ot is the 2200 m/s fluence, $\hat{\sigma}$ is known as the Westcott cross section for the isotope ^{30}Si, and is given by:

$$\hat{\sigma} = \sigma_o \left(G_{th}g + G_r S_o r \sqrt{T/T_o}\right) \qquad (2)$$

where c_o is the 2200 m/s capture cross section for the isotope ^{30}Si; G_{th}, G_r are dimensionless self-shielding factors for thermal neutrons and epithermal neutrons in silicon ingots, respectively, ($G_{th}=G_r=1$); g is a measure of the departure of the ^{30}Si capture cross section from 1/v dependence averaged over the Maxwellian component of the neutron spectrum, (g=1); S_o is a resonance parameter of ^{30}Si ($S_o = 0.613$)[4]; $r \sqrt{T/T_o}$ is an epithermal index of the neutron spectrum of the irradiation position. Evidently, both thermal and epithermal neutron contributions to the generation rate of ^{31}P are considered simultaneously. The irradiation fluence is determined by a cobalt foil with a Westcott cross section, g = 1.0, S_o = 1.75, σ_o = 37.5 b[5] and Gh = 0.985, Gr = 0.74. Based on the relationship between resistivity and carrier concentration $\rho = (C_p\mu_d e)^{-1}$, the equations between the fluence nv_ot, the target resistivity ρ_t and the initial resistivity $\rho_n{}^o$ or $\rho_p{}^o$ can be derived:

$$nv_ot = \frac{1}{K} \left(\frac{1}{\rho_t} - \frac{1}{\rho_n{}^o}\right) \text{ for initial n-type materials} \qquad (3)$$

$$nv_ot = \frac{1}{K} \left(\frac{1}{\rho_t} + \frac{\mu_d}{\rho_p{}^o \mu_a}\right) \text{ for initial p-type materials} \qquad (4)$$

$$K = N_o \hat{\sigma} \mu_d e \qquad (5)$$

where K is called the doping factor. Substituting the value of ρ_t, ρ^o measured by a four-point probe and the value of nv_ot measured by cobalt foils in equation (3) or (4), the K value can be determined experimentally. Carrying out the statistical average over a great number of results with the target resistivity of 10 Ω-cm to 300 Ω-cm, we obtained K = $(4.1 \pm 0.1) \times 10^{-20}$.

In equation (5) N_o = 1.544×10^{21} cm^{-3} and e = 1.602×10^{-19} coulombs are the experimental values. Substituting values for N_o, e and K into equation (5), we can obtain $\hat{\sigma}\mu_d$ = 1.66×10^{-22}. Considering $r \sqrt{T/T_o}$ = 0.024 and $\hat{\sigma}$ = 1.015 σ_o in the HWRR irradiation position of silicon ingots, $\sigma_o\mu_d$ = 1.63×10^{-22} which is in agreement with the results reported at the last conference in Copenhagen[6,7].

Calibration of Self-Powered Detectors. In practice the irradia-

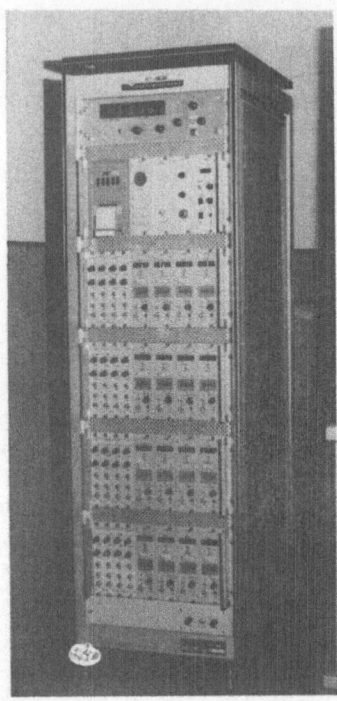

Fig. 6.   Current integrator.

tion fluence is given by the activation of the cobalt or cobalt-
aluminum alloy foil attached to each can. This fluence can be
used for determining the doping factor and calibrating the self-
powered detectors.

The accuracy of NTD largely depends upon the accuracy and
stability of the self-powered detectors. The operation showed
that our self-powered detectors and integrators (see figure 6)
have good linear characteristics and stability with accuracy within ±2 percent in a reactor operation cycle.

The main factors that influence the calibration accuracy are
the thermal flux distortion caused by the cans, the movement of
the control rods, and the depletion of the self-powered detectors.
Experimental studies of these factors were made in detail.

In the case of the HWRR, the case and detectors are all in
the dry tubes. When silicon ingots with different diameters are
put into the facility, the thermal flux depression is 3-6 percent.
However, the current of self-powered detectors are also reduced in

the same way, so that the variation of the calibration value is less than one percent. The experiments indicated that the variation of the calibration value is about 2-4 percent due to the movement of the control rods. The calibration value should be slightly modified according to the height of the control rods in the different operation cycles. In a cycle the depletion of self-powered detectors is also very small. Therefore, the discrepancy between the predetermined and real fluence values can be controlled within ±5 percent.

In the case of the SPR, the variation of thermal flux is quite large and is dependent on the size of ingots and the loading condition in the neighbouring tubes. However, the self-powered detectors could not fully respond to these variations. Experimentally, having determined these factors, correction curves were obtained.

The present discrepancy between the predetermined and real fluence is less than ±5 percent for 90 percent of the silicon ingots.

Control of Irradiation Uniformity

The radial uniformity of NTD is ensured by means of rotating the irradiation cans. Radial distribution measurements of ^{31}Si across the irradiated silicon slices with a silicon-lithium detector show that the radial nonuniformity of NTD is less than two percent for silicon ingots with a diameter less than 75 mm.

For flattening the axial flux, we use the method of "upside down" or "position exchange" rather than neutron screen so as to keep the irradiation capacity[7]. As shown in figure 7, can No. 1 is located in the flat-flux region of HWRR core. The axial non-uniformity over the length of 300 mm is less than ±4 percent. Can No. 2 is located in the linear-flux region and will be turned upside down at one-half of the predetermined fluence. Thus, the axial nonuniformity over the length of 300 mm is less than ±2 percent.

In the case of the SPR, an irradiation basket with two silicon ingots, each 150 mm long, is placed in the core in such a manner that its midpoint coincides with the point of maximum axial neutron flux. At one-half of predetermined fluence an "upside down" or "position exchange" operation is necessary. It ensures an axial nonuniformity of less than ±3.5 percent; otherwise, it is about ±10 percent. See figure 8.

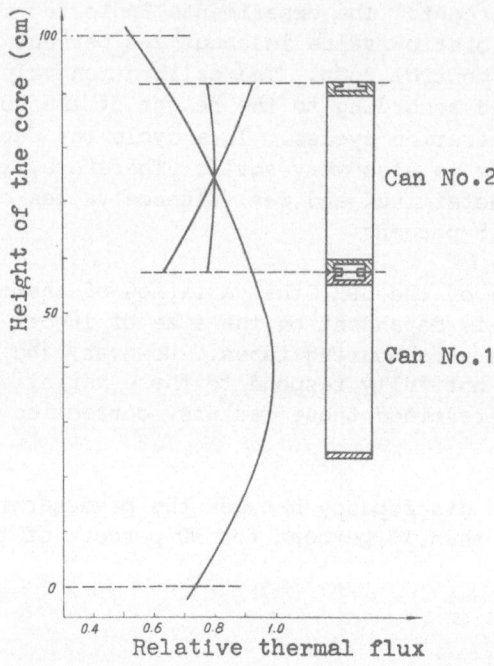

Fig. 7.   Typical thermal-flux profile in the HWRR.

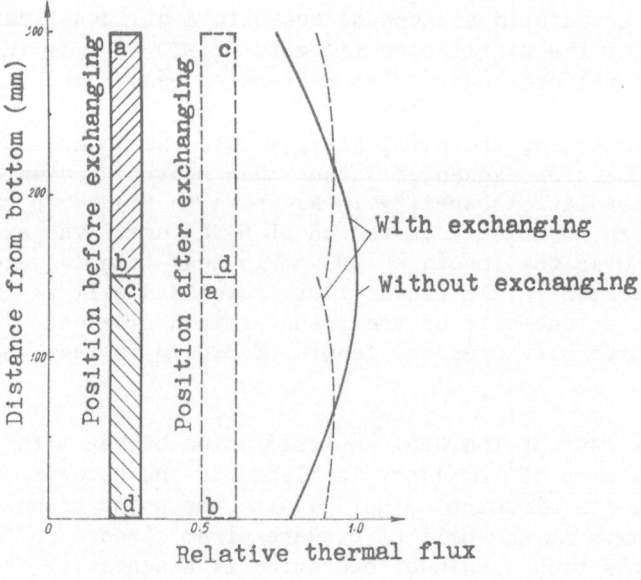

Fig. 8.   Typical thermal-flux profile in the SPR.

Cleaning and Residual Radioactivity Monitoring

Cleaning of irradiated silicon ingots is quite easy as long as the surfaces of silicon ingots are kept clean before putting them into the cans. In general, the silicon ingots are immersed in a detergent in an ultrasonic bath. A few ingots must be etched by a mixture of nitric and hydrofluoric acids with the proportion of 1:10.

According to the Chinese regulation on radioactivity protection and international standard, our exempt level of irradiated silicon ingot is defined as follows:

1. ^{32}P specific β-activity is less than 5×10^{-4} $\mu Ci/g$;
2. The total specific activity from all γ emitting nuclides is less than 10^{-5} $\mu Ci/g$;
3. The removable surface contamination is less than 10^{-4} $\mu Ci/cm^2$.

Every silicon ingot is monitored by a special device to ensure it is below the exempt level. The silicon ingots are dispatched with an exempt certificate.

IRRADIATION RESULTS AND CAPACITY

NTD silicon with target resistivity from 10 Ω-cm to 10000 Ω-cm has been produced. Devices such as high power rectifiers, thyristors, power diodes, transistors, C-MOS integrated circuits, and semiconductor detectors were fabricated with these materials. The quality of all these products was improved and the cost was reduced.

The two reactors have a large irradiation capacity. For a 50 Ω-cm target resistivity, it is about 2 tons per year for the SPR, and 8 tons per year for the HWRR. There is still a greater potential capacity available. In addition to meeting the domestic needs, we would like to provide an irradiation service for other countries.

REFERENCES

1. Lu Cuengang and Li Yaoxin, Neutron Transmutation Doped Silicon, ed. by Jens Guldberg, p. 273 (1981).
2. Ma Fubang, etc., Chinese Journal of Nuclear Science and Engineering, Vol. 1, No. 1, p. 20 (1981).
3. K. H. Beckurts and K. Wirtz, Neutron Physics, p. 279 (1974).
4. S. F. Mughabghab and D. I. Garber, Neutron Cross Section, Vol. 1, Resonance Parameters, BNL 325, 3rd ed. (1973).

5. Handbook on Nuclear Activation Cross-Section, Tech. Reports
    Series No. 156, p. 17 (IAEA Vienna, 1974).
6. K. Heydorn and Kirsten Andresen, Neutron Transmutation Doped
    Silicon, ed. by Jens Guldberg, p. 193 (1981).
7. N. W. Crick, Neutron Transmutation Doped Silicon, ed. by Jens
    Guldberg, p. 211 (1981).
8. S. L. Gunn, J. M. Meese, and D. M. Alger, Neutron Transmuta-
    tion Doping in Semiconductor, ed. by J. M. Meese, p. 197
    (1979).

MEASUREMENTS OF THE GAMMA ABUNDANCE OF SILICON-31

E. J. Parma, Jr. and R. R. Hart

Department of Nuclear Engineering
Texas A&M University
College Station, Texas 77843

ABSTRACT

Direct determination of the induced ^{31}P concentrations in neutron transmutation doped silicon may be obtained by measurements of absolute activities of ^{31}Si by detection of the 1.266 MeV gamma-rays that are emitted, provided the gamma abundance is known. In previous work it was inferred that the gamma abundance of ^{31}Si is significantly lower than the accepted value of 7 x 10^{-4}. To confirm this result a 4π proportional counter and a calibrated Ge(Li) detector were used to measure both the absolute beta and gamma activities, respectively, of thin silicon samples. The resultant gamma abundance of ^{31}Si was found to be 5.9 x 10^{-4} ± 5%.

INTRODUCTION

The effects of neutron transmutation doping of silicon are dependent on the absolute concentrations of ^{31}P that are produced. These concentrations may be determined from measurements of the absolute activity of ^{31}Si as the ^{31}Si atoms decay to ^{31}P.[1] Previous investigations have shown that ^{31}Si decays with a half-life of 2.62 hours to the ground state of ^{31}P by means of a single beta emission, E = 1.49 MeV, or by a beta emission, E = 0.22 MeV, accompanied by a gamma-ray of 1.266 MeV.[2] Provided the gamma abundance is known, ^{31}Si activities may be readily measured by detection of the 1.266 MeV gamma-rays.

In earlier work by Lyon and Manning[3] the gamma abundance of ^{31}Si was found to be 7×10^{-4}. However, based on recent neutron transmutation doping studies of silicon the gamma abundance of ^{31}Si was inferred to be $5.6 \times 10^{-4} \pm 10\%$.[1] The objective of the present work was to resolve this discrepancy by a more precise determination of the gamma abundance of ^{31}Si.

EXPERIMENTAL

The basic approach used in this work was to use a 4π proportional counter and a calibrated Ge(Li) detector to measure both the absolute beta and gamma activities, respectively, of thin silicon samples. Since a beta particle is emitted in both decay modes of ^{31}Si, the total disintegration rate of ^{31}Si atoms is equal to the beta activity. Hence, the ratio of the gamma to beta activities gives the gamma abundance of ^{31}Si.

Silicon wafers, each weighing approximately 2 mg and having a diameter of 0.64 cm, were irradiated at the Texas A&M University Research Reactor. The production of ^{31}Si was based on the ^{30}Si$(n,\gamma)^{31}$Si reaction. Each wafer was irradiated for 2 hours in a thermal neutron flux of approximately 5×10^{12} n/cm^2-sec and then chemically etched to reduce surface contamination.

A gamma spectrum of each activated sample was obtained using a Ge(Li) detector coupled with a computer-based 4096 channel analyzer. The saturated gamma activity of each silicon wafer was determined using the equation:

$$A_{\infty\gamma} = \frac{L_\gamma C_\gamma}{\varepsilon_\gamma}$$

where:

L_γ = $\lambda(tr/tc)e^{\lambda tw}/ (1-e^{-\lambda ta})(1-e^{-\lambda tr})$

C_γ = observed counts minus background for the 1.266 MeV gamma-ray peak,

ε_γ = overall efficiency of the detection system,

λ = decay constant of ^{31}Si,

tr = real counting time,

tc = live counting time,

tw = wait time,

ta = activation time.

Equation (1) is exact when the fractional dead time of the system
is linear throughout the counting interval. The overall efficiency includes the detector efficiency, the transmission factor
due to self absorption, and the geometry factor.

A detector efficiency curve was determined using a National
Bureau of Standards point source placed at the same location
above the detector as the silicon wafers. A typical curve of
this type is shown in Reference 1. For the gamma-ray energy of
^{31}Si, 1.266 MeV, the detector efficiency was determined to be
0.0126 ± 3%. Based on a calculated self-absorption coefficient
of 0.33 cm^{-1} the transmission factor was estimated to be greater
than 0.999. Since the diameters of the silicon samples and the
NBS point source were approximately equal, the geometry factor
was taken as 1.00. This estimate was substantiated using a
mm-sized ^{60}Co source. The observed count rate decreased by only
0.5% as the source position was varied over the radial position
of a silicon wafer.

After a gamma spectrum was obtained for an activated silicon
wafer, the beta activity, i.e., total activity, was measured for
the same wafer using a 4π counter. The 4π counter consisted of
two hemispherical chambers separated by a thin aluminum-coated
mylar planchet. Each chamber had a separate charge-collection
loop. During normal operation, the two collecting loops were
connected in parallel. Thus, a beta particle could cause ionization in either or both halves of the detector and cause a single
event to be recorded.

After amplification the voltage pulses were counted by a
fast scaler which incorporated a lower-level discriminator. The
pulse-height distribution was also determined using a multichannel analyzer and the results are shown in Figure 1. The amplification of the system was adjusted so as to minimize the number of
low-energy pulses that were below the 50 mV voltage thresholds of
the scaler and multichannel analyzer while also requiring that
less than 1% of the high-energy pulses exceeded the 10 V maximum
of the scaler. Although clipped pulses were counted, they tended
to increase the resolving time of the system.

To verify that essentially all pulses were counted, the integral count rate as a function of the lower-level discriminator
voltage was measured and the results are shown in Figure 2. It
can be seen that the count rate is constant to within 1% at discriminator settings less than 0.1 V. The discriminator was then
fixed at the nominal setting of 0 V which corresponded to a
threshold of 50 mV. Electronic noise was sufficiently smaller in
amplitude than this threshold so as to negligibly contributed to
the observed background count rate of 70 cpm. The estimated beta

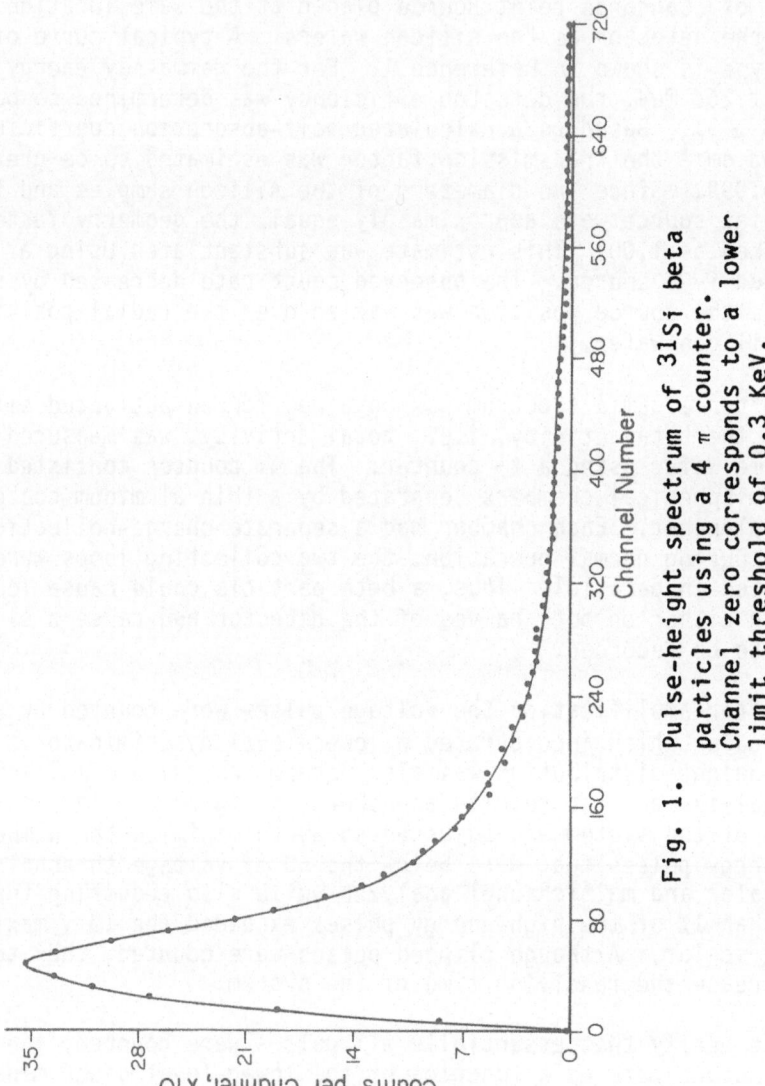

Fig. 1. Pulse-height spectrum of 31Si beta
particles using a 4 π counter.
Channel zero corresponds to a lower
limit threshold of 0.3 KeV.

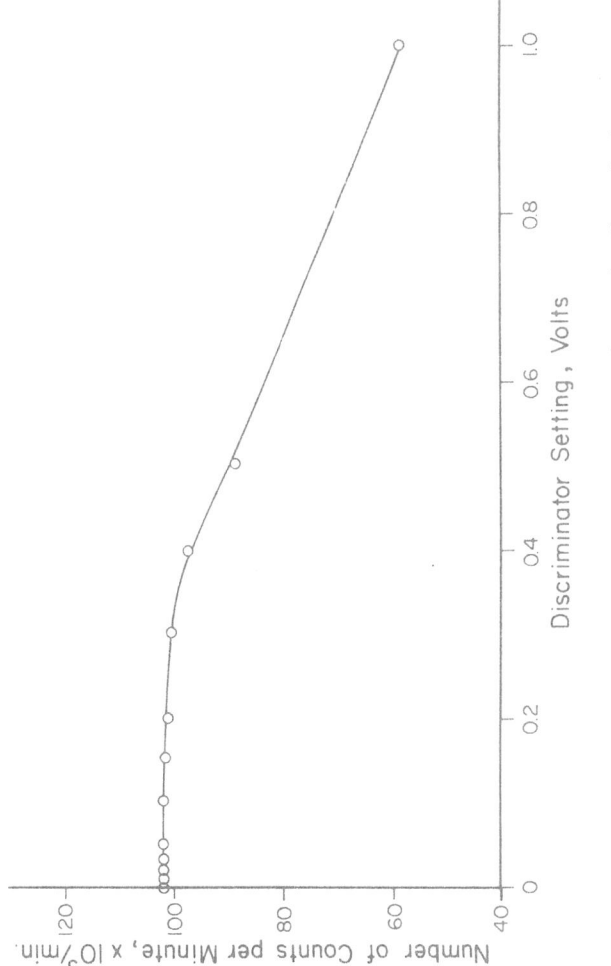

Fig. 2. Integral count rate as a function of discriminator setting for the pulse height spectrum of Fig. 1. A zero voltage setting corresponded to an actual voltage of 0.05 volts. The error bars are not shown, but include less than the circle representing each point.

activity was determined using the equation:

$$A_{\infty\beta} = \frac{L_\beta \, C_\beta}{\varepsilon_\beta} \qquad\qquad (2)$$

where:

$L_\beta = \lambda e^{\lambda tw}/(1-e^{-\lambda ta})(1-e^{-\lambda tr}),$

C_β = observed counts minus background,

ε_β = overall efficiency as defined below.

The overall efficiency included the intrinsic detector efficiency, the transmission factor of the mylar planchet, the dead time correction factor, and the transmission factor due to self-absorbtion.

The intrinsic efficiency is defined as the fraction of beta particles emitted that produce pulses greater than the threshold energy. The primary concern is that the betas emitted with energies near the endpoint energy of 1.49 MeV as well as low energy betas may not cause sufficient ionization over the 3.4 cm radius of the detector to produce pulses greater than the threshold. Using an [55]Fe source, which emits 5.9 KeV x-rays, the threshold energy was determined to be 0.3 KeV. The fraction of beta particles which deposit energies below the threshold was appoximated by using a theoretical beta spectrum[4] for [31]Si and stopping power data for the P-10 gas used in the counter.[5] It was estimated that less than 0.1% of the betas emitted would not be detected. The intrinsic efficiency was therefore estimated to be 0.999.

This estimate was substantiated using a thin ^{32}P standard (Emax = 1.710 MeV) that was prepared from a solution of known specific activity. The solution was obtained and calibrated commercially. Activity measurements agreed to within ±2% of the calibrated value which was specified to within ±5%.

The transmission factor of the mylar planchet was determined to be 1.00 ± 0.3% by measurements of the difference in count rates between the two halves of the 4π counter. The dead time correction factor, f_d, is given by

$$f_d = 1 - n\rho$$

where:

n   = observed count rate,

ρ = resolving time.

From roll-off at high count rates, the resolving time was deter-

mined to be $18\,\mu$ sec. Thus, f_d is greater than 0.99 if the observed count rate is less than 30,000 cpm.

Self-absorption of the [31]Si beta particles within the silicon samples was examined using a stack of up to four silicon
wafers. The specific activity was measured as each wafer was
placed directly on top of the preceding one. Extrapolation to
full transmission at zero thickness gives the fraction transmitted as a function of thickness. The results are shown in Figure
3. The solid line is a least squares fit of the data to the
equation:

$$f_s = \frac{1-e^{-\mu X}}{\mu X}$$

where:

f_s = transmission factor,

μ = self-absorption coefficient,

x = sample thickness.

From this fit μ was determined to be $0.009 \pm 17\%$ cm^2/mg. This
result is consistent with a semi-empirical value of 0.0096 cm^2/mg
for an aluminum absorber.[6] Also from Figure 3 the transmission
factor for one of the present silicon wafers (thickness = 7.3
mg/cm^2) is $0.97 \pm 2.4\%$.

RESULTS

A typical beta decay curve of an activated silicon wafer
using the 4π counter is shown in Figures 4 and 5 on different
time scales. Roll-off at very high count rates is caused by system dead time. This decay curve is seen to initally decrease
exponentially with a 2.6 hour half-life which is characteristic
of [31]Si. After 40 hours a long-lived component having a half-
life of 2.7 days becomes evident. As confirmed by gamma spectroscopy, this component is [198]Au (half-life = 2.695 days) but
contributed less than 0.1% of the total activity at a reference
time of zero. A weighted least squares fit of the data was performed using the CLSQ computer program.[7] This program incorpor-
ates the decay factors of eqn. (2) as well as the dead time correction factor for each point to give the intercepts at zero-ref-
erence time and the half-lives of the two components. The half-
life of [31]Si was found to be 2.62 ± 0.01 hours in agreement with
Reference 2. The wait time correction factor and the remaining
efficiency factors of eqn. (2) were then used to find the saturated beta activity.

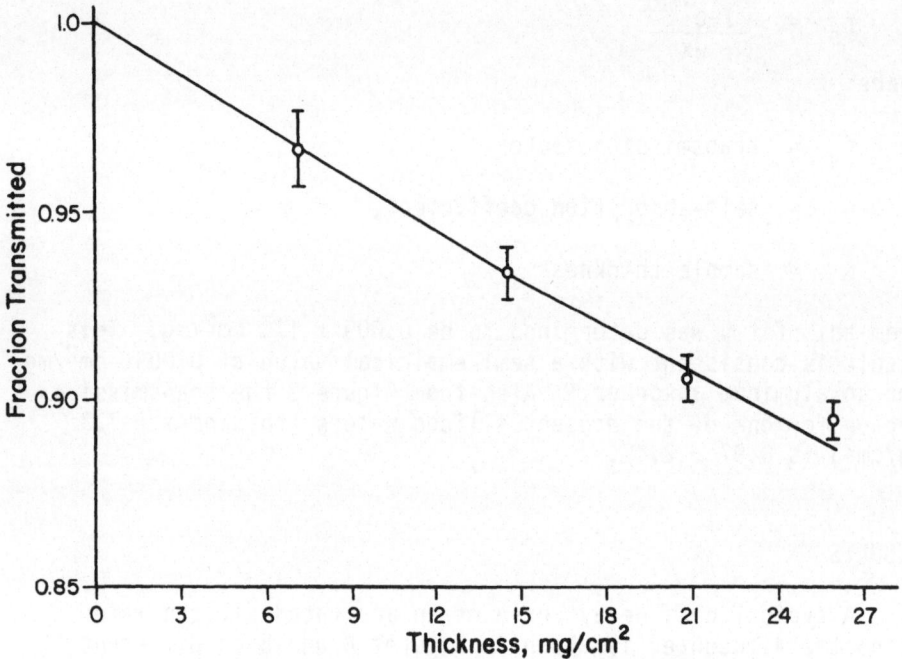

Fig. 3. The fraction of ^{31}Si beta particles transmitted as
a function of silicon thickness. For a single
silicon wafer used in this measurement (7.3
mg/cm^2), the transmission factor is 0.97 ± 2.4%

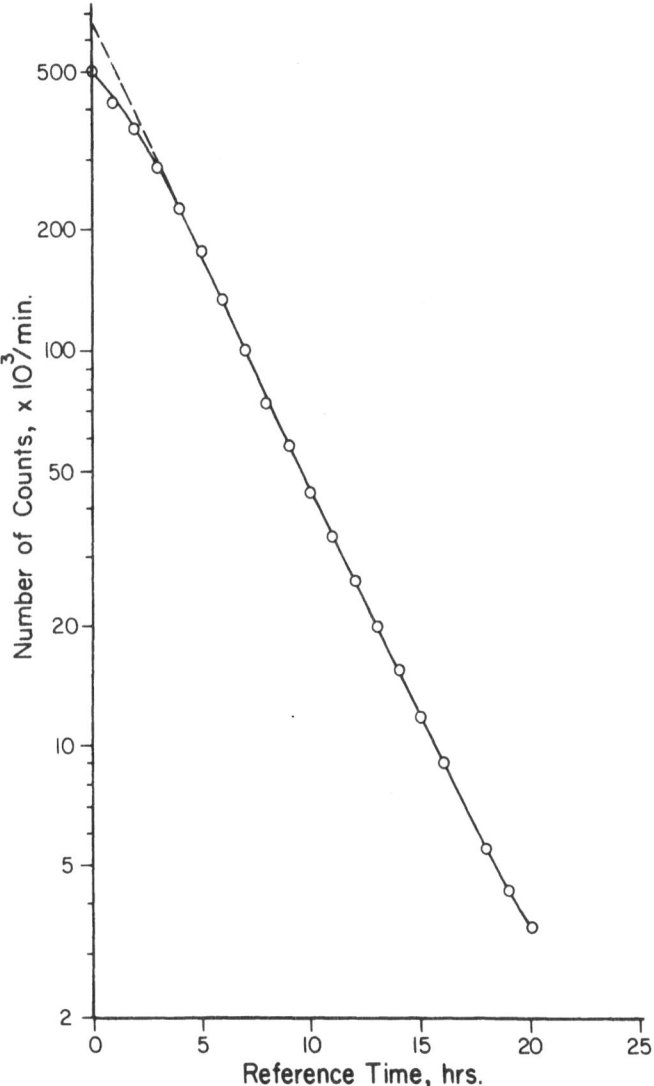

Fig. 4. A typical beta decay curve of an activated silicon
wafer using the 4π counter. Roll off is observed
at high count rates due to system dead time. Error
bars are smaller than the size of the circles
representing each point.

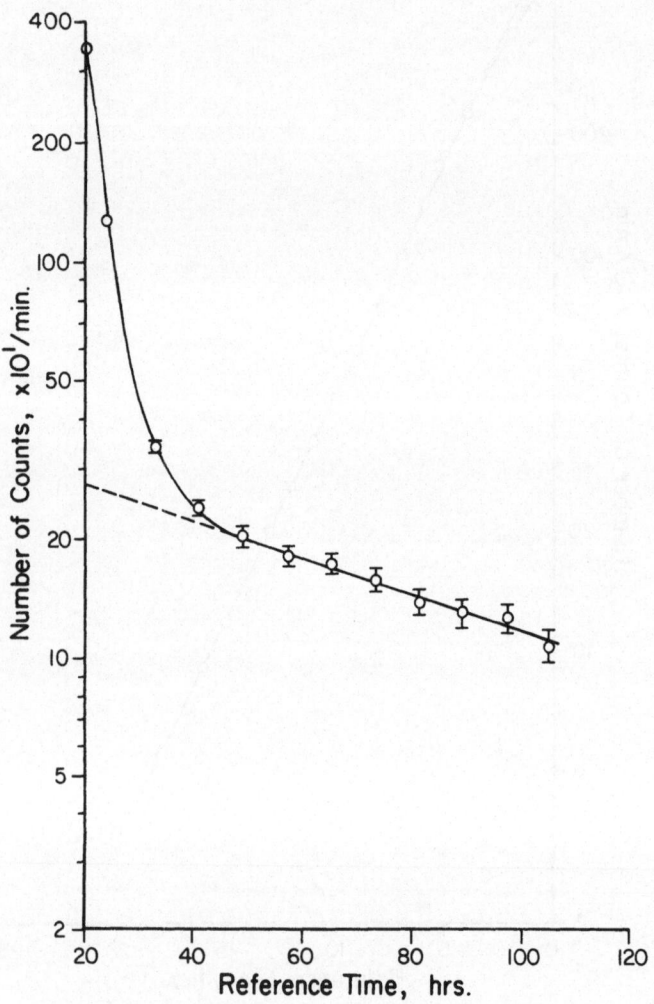

Fig. 5. A continuation of the beta decay curve of
Fig. 4 showing a long-lived component that
was identified to be ^{198}Au.

The estimated gamma activity, which was determined using eqn. (1), was divided by the saturated beta activity to give the gamma abundance of ^{31}Si. Four seperate runs were made. The results varied from 5.7×10^{-4} to 5.9×10^{-4}. Averaging the results the gamma abundance of ^{31}Si is $5.9 \times 10^{-4} \pm 5\%$. The uncertainty in the final result is a conservative estimate which includes statistical errors as well as experimental errors of the various correction terms.

CONCLUSIONS

The present result for the gamma abundance of ^{31}Si agrees within experimental error with the value that was inferred in Reference 1. Consequently, the conclusion of Reference 1 is substantiated, i.e., ^{31}P produced by neutron transmutation doping of silicon is completely electrically active following 850 °C anneals of float-zoned silicon. Furthermore, this corrected value for the gamma abundance of ^{31}Si permits accurate measurements of ^{31}Si activities by detection of the 1.266 MeV gamma rays. Such measurements permit direct calibrations of ^{31}P production in various reactor facilities.

ACKNOWLEDGEMENTS

The authors express their appreciation to A. Lee for preparation of the silicon samples and to O. J. Marsh and M. H. Young, Hughes Research Laboratories, for their support of this work.

REFERENCES

1. R. R. Hart, L. D. Albert, N. G. Skinner, M. H. Young, R. Baron, and O. J. Marsh, in Neutron Transmutation Doping in Semiconductors, Jon M. Messe, Ed. (Plenum, New York, 1979), p. 345.
2. E. Browne, J. M. Dairiki, and R. E. Doebler, Table of Isotopes (John Wiley and Sons, Inc., New York, 1978).
3. W. S. Lyon and J. J. Manning, Phys. Rev. 93, 501 (1954).
4. I. Kaplan, Nuclear Physics (Addison-Wesley Publishing Co., Inc., Reading, Mass., 1962).
5. L. Pages, E. Bertel, H. Joffre, and L. Sklavenitis, Atomic Data 4, 1 (1972).
6. G. I. Gleason, J. D. Taylor, and D. L. Tabern, Nucleonics 8, 12 (May 1951).
7. J. B. Cumming, BNL-6470 (1962).

EXPERIENCES WITH THE NORWEGIAN RESEARCH REACTOR

JEEP II IN NEUTRON TRANSMUTATION DOPING

N. Kaltenborn
Institute for Energy Technology, Kjeller, Norway
And
O. Malmros
Topsil A/S, Frederikssund, Denmark

SUMMARY

The Norwegian research reactor JEEP II, situated at Kjeller, 25 km northeast of Oslo, has been in use for neutron doping of silicon since 1975. Ordinary radionuclide production irradiation facilities are used, with some modifications of the loading equipment. The neutron flux range for this type of irradiation is $0.5-1.4 \times 10^{13}$ n/cm^2 sec, and the maximum diameter of silicon ingots that can be handled is 78 mm.

The available effective volume for silicon irradiation is about 32 litres. For neutron flux variations within ± 5 percent this volume is reduced to about 13 litres, corresponding to a silicon charge of 30 kg.

The reactor must be shut down to permit insertion and removal of the silicon charge. However, a practical system has been developed in cooperation with Topsil A/S, Denmark, and the Institute for Energy Technology, Norway (the owner of the reactor) to make the best utilization of the reactor facility for its intended purposes.

1. INTRODUCTION

The development of a neutron doping service utilizing this tank-type heavy-water reactor has been undertaken in cooperation with the owner of the reactor,

the Norwegian Institute for Energy Technology, and the
Danish company Topsil A/S. The reactor was originally
built for other purposes. It was completed in 1966, and
some modifications of the neutron irradiation equipment
were necessary to adapt it to neutron transmutation
doping of silicon ingots. Irradiations for this purpose
have been carried out since 1975.

2. DESCRIPTION OF THE REACTOR FACILITY

    The construction and building of the reactor was
carried out during the years 1961-1966 by the Institute
for Energy Technology's staff in cooperation with a
Norwegian consultant company in nuclear services,
Noratom A/S. The reactor was completed for the sum of
3.3 million US dollars, and is situated in eastern
Norway, about 25 km east of Oslo.

    No provision had been made initially for the
irradiation of relatively large pieces of material like
silicon ingots. The primary utilization of the reactor
has been for experiments in solid-state physics and
irradiation of materials for production of radioisotopes
for use in medicine, industry and research (Figs. 1-2).

    The fuel of the reactor is made of 3.5 percent
enriched uranium dioxide pellets, contained in vertically
positioned aluminium-canned fuel elements. The aluminium
tank contains about 5.4 tons of heavy water, and the
working temperature is about $40^{\circ}C$. The fuel cycles have
a duration of about 5.5 years, divided in four parts;
the first 1.5 years the reactor contains 16 fuel elements,
systematically increasing to 19 elements in the last
quarter period. The thermal neutron flux is kept as
closely as possible to the same average level throughout
the cycle. The maximum neutron flux level is approximately 2×10^{13} n/cm^2 sec.

    All handling of materials for irradiation is
undertaken from the top of the reactor block, utilizing
vertical channels, while neutron beams for physics
research work are taken from the reactor tank
horizontally. Except for small samples irradiated in a
pneumatical "rabbit" transfer system, changing materials
for irradiation requires a shut-down of the operation
of the reactor.

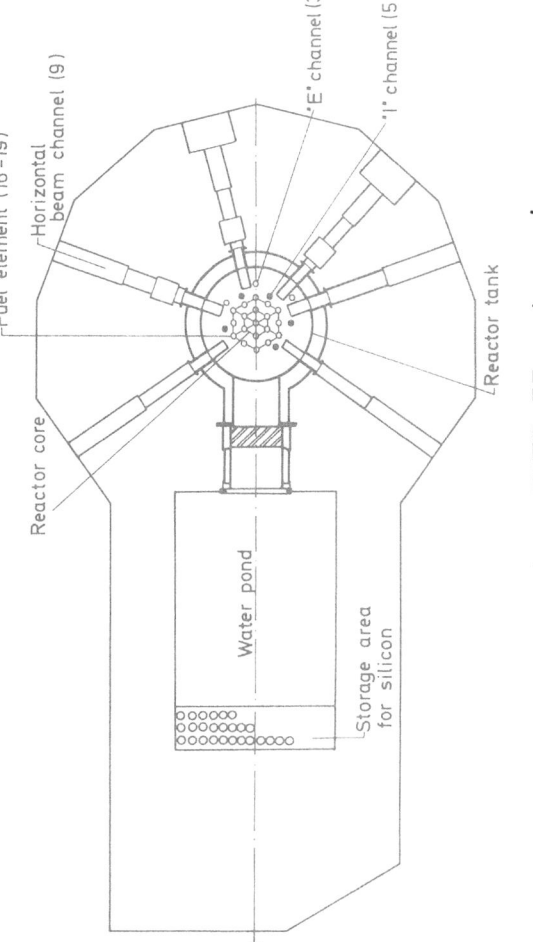

Fig. 1. The reactor JEEP II - top view

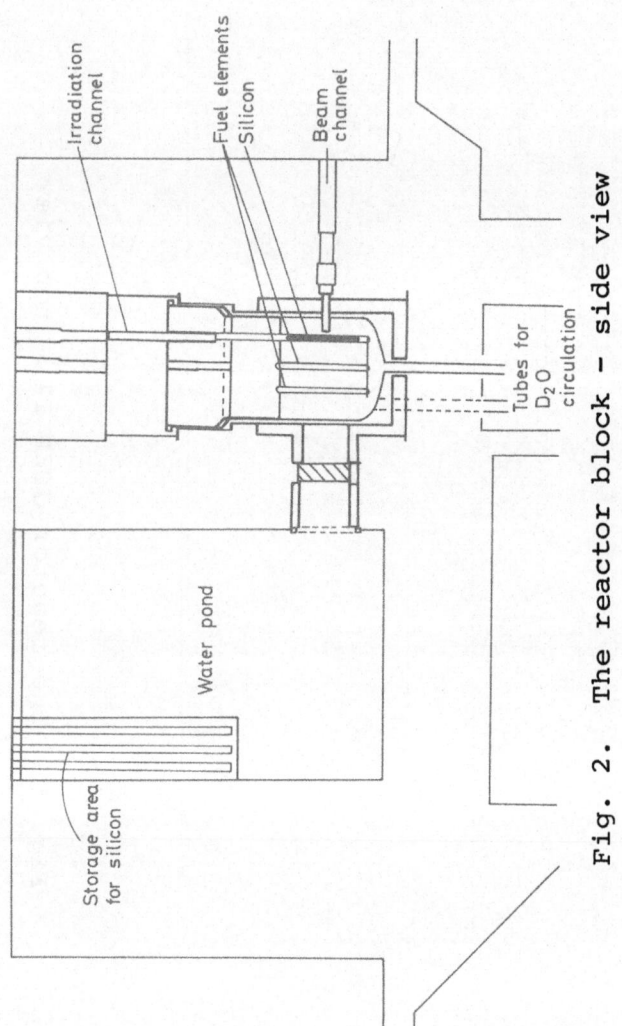

Fig. 2. The reactor block – side view

3. IRRADIATION FACILITIES

The reactor has a total of eleven vertical channels available for irradiation. Rack tubes of aluminum containing the material to be irradiated are placed into these vertical channels. An irradiation channel with rack tube (silicon container) is shown in Fig. 3. The diameter of the racks is about 80 mm, and the effective height is 850 mm. For the irradiations performed for production of radioactive materials aluminum cans of dimensions 30 and 230 cm^3 respectively are used.

The eleven vertical channels have been grouped into three various types, with respect to the distance from the reactor core center, cfr. Fig. 4.

a) Three low-flux irradiation channels situated in the reflector part of the reactor tank (E2, E3 and E4).

b) Five isotope production channels situated just outside the core (I2, I3, I4, I5 and I6).

c) Three "fuel element positions" available for irradiation: 36, 49 and 52

4. ADAPTION OF THE IRRADIATION FACILITIES TO NEUTRON DOPING OF SILICON

Irradiation of silicon was started during the second half of 1975, utilizing the two outermost irradiation channels, E3 and E4. The silicon ingots were placed inside aluminum cans of dimensions 54 mm in diameter and 115 mm in height. The neutron fluxes were measured along the axis of the channels by means of irradiating small cobalt sources in a prescribed pattern, and the same system is still in use. The radial flux gradient was found to be unacceptably high, about 6 percent. This was overcome by installing motor-driven turning equipment on the complete channel installation. After half the required irradiation time has elapsed, the channel unit is simply turned 180 degrees. This simple method has been used successfully all the time, and the turning operation is now part of the reactor operation schedule.

The limitation of ingot size to 2 inches in diameter and to about 115 mm in height was not acceptable to the

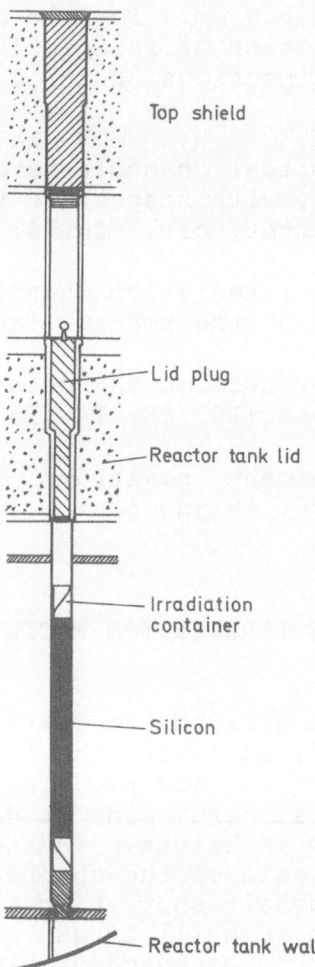

Top shield

Lid plug

Reactor tank lid

Irradiation
container

Silicon

Reactor tank wall

Fig. 3.   Irradiation channel

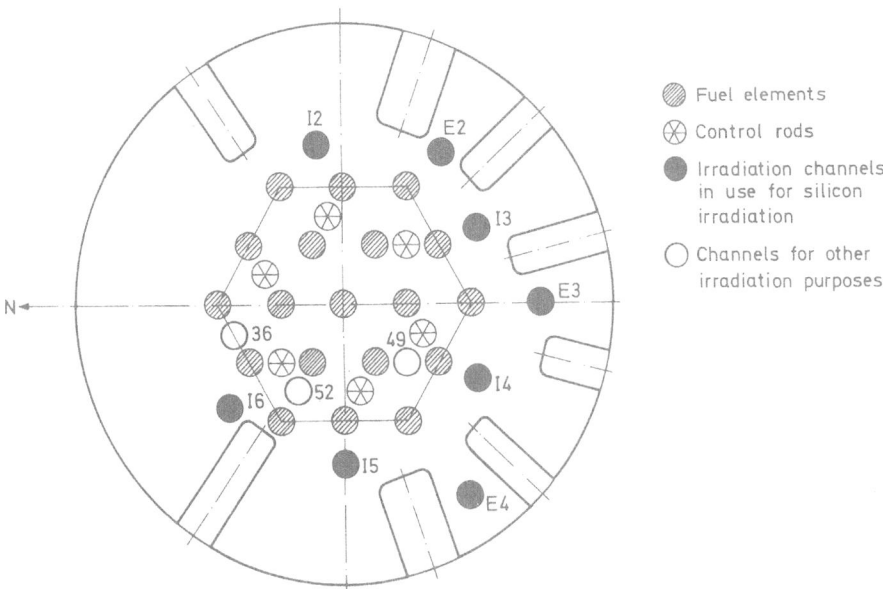

Fig. 4.   Reactor core and reflector arrangement

silicon producer, Topsil A/S, in the long run. Open
tubes to be loaded from the bottom were therefore
constructed, and with small modifications, the same
type of irradiation tubes are in use at the present time.

During 1976 two more channels were used for silicon
irradiation, E2 and I4, and at the present time altogether
eight of them have 180 degrees' turning facilities. The
in-core channels have been considered less useful for
neutron doping, since the ratio of fast neutron flux
to thermal neutron flux is rather high:

	Distance from reactor center mm	Thermal flux 0-0.625 eV (x 10^{13}_{2} n/cm^2 sec)	Fast flux 0.625 eV- 10 MeV (x 10^{13}_{2} n/cm^2 sec)	Ratio of fast to thermal neutron flux
E channels	560-700	0.75	0.1	0.13
I channels	480	1.4	0.4	0.29
C channels	260-300	2.0	1.6	0.80

They are, however, useful for radionuclide production
for Scandinavian hospitals, research laboratories, and
industry.

The maximum size of ingots for irradiation is now
78 mm diameter and 840 mm length. With a reactor core
height of 900 mm, however, the flux drop is rather
significant both in the upper and the lower part of this
maximum length. Varying with the requirements for doping
accuracy, normally from 360 to 500 mm of the available
height is utilized, i.e., compared to a maximum practical
load for each channel per irradiation of 10.2 kg Si the
usual amounts range from 3 to about 6 kg.

The use of steel flux flatteners has been discussed,
but not implemented. Positive effects would be a better
utilization of the channel length, and within some limits
also better doping accuracy for shorter pieces of
material. On the negative side is the loss of diameter
of irradiation tubes, since the reactor construction
will not permit flux flatteners to be placed outside
the irradiation channels. One has also to contend with
some loss of irradiation capacity and of flexibility in
arranging the weekly planned irradiation program. The
plans at present are to test the use of the flux
flatteners in one or two of the least feasible of the
channels based on axial flux specifications.

5.   IRRADIATION PROGRAMMING AND MATERIAL MANAGEMENT

In the programming of the silicon irradiation
service the following characteristics of the reactor
have to be considered:

a)   The reactor operation has to be shut down before
     changing the irradiation material load.

b)   The various irradiation positions in use have
     rather different neutron fluxes.

These facts, combined with the various target
resistivity specifications for each ingot, make the
irradiation programming appear to be a rather complicated
affair. However, by careful organization it turns out
to work very well.

Neutron flux data are measured four times a year,
usually after stabilizing the reactor, following a major
shut-down period, which takes place in April, July, and
December every year, altogether about ten weeks. The
remaining 42 weeks the reactor is available for irradiation, and shut-down periods for changing of the
load can be scheduled at any time, to make the desired
neutron doses to the silicon ingots. To stop the reactor,
reload it, and start it takes about one hour.

On the basis of the last set of flux data a chart
of the average neutron doses corresponding to various
ingot lengths is worked out, and from these data the
correct time of irradiation for each single ingot or set
of ingots is calculated. The length of the irradiation
period thus varies from channel to channel, and there
is a factor of about 2.5 between the shortest and the
longest period for a specified dose.

After the ingot groups have been distributed to the
various irradiation positions and the correct irradiation
periods have been calculated, a plan for the following
week is worked out. A normal week is divided into 25
to 30 irradiation periods of various lengths. In this
planning factors outside the silicon irradiation service
have to be considered: The reactor physics and solid-
state physics groups have their demand for certain days
with long working periods with no stops, the activation
analysis group prefers all shut-downs to be outside the
working hours, and the reactor operations group wants
to avoid elaborate operations during the night. A

written program on a weekly basis is distributed, of
course, but the crucial point of it is the length of
the operation periods and not the exact times of the
stops and starts. Digital displays are therefore used
in the various laboratories around the reactor center in
which the number of the irradiation period, the time
left to the next stop, and the planned time for the next
start and stop are shown.

The handling of the material is undertaken as
follows: A special storage area has been prepared,
directly accessible from the top of the reactor block.
Silicon ingots are loaded into irradiation tubes and
stored in numbered positions in part A of the storage
area. As a part of the reactor operation the specified
tubes are put into the reactor in the prescribed
channels, taken out after the correct irradiation period,
and placed in part B of the storage area, from where
the ingots are unloaded after a certain decay time.
Written instructions are given to the reactor operations
group for each stop period, with orders of which channels
to unload and reload, correct hour for 180 degrees'
turning of all channels, etc.

6.   VARIATIONS IN THE NEUTRON FLUX

The requirement of the silicon producer is to have
his material irradiated in as large units as possible
with the neutron flux variation over that volume of
material as small as possible. No reactor, constructed
and operated mainly for other purposes, can fulfill
this requirement completely. Constructional additions
have led to an improvement in the flux variation. The
rotation of the irradiation tubes, the use of metal
absorbers for the reduction of the highest flux levels,
and the turning of the irradiation tube upside down for
the second half of the irradiation period are methods
that have been reported at earlier conferences. Such
devices are generally more easily adaptable to reactors
of the swimming-pool type than to the tank-type.

As mentioned, radial flux variations are almost
completely eliminated in the Kjeller reactor by means
of simple 180 degrees' turning equipment. The axial
variations are more complicated to handle, as the
reactor construction will not allow any upside-down
turning of the irradiation tubes to be made.

During the fuel cycle of about 5.5 years the neutron flux is altered significantly, but rather slowly. Most of the changes occur in steps because of the addition of three fresh fuel elements during the cycle. In addition the gradual burn-up of the fuel causes the control rods to move upwards, thus increasing the flux in the top of the channels and reducing it in the lower end. Fig. 5 shows the variation of the flux distribution in the position E2 over a period of 4.25 years. Notice that the average flux is gradually reduced, and also shifts from the upper part to the lower part of the irradiation channel.

In practice it has been found adequate to perform calibration measurements by means of small cobalt discs four times a year. Fig. 6 shows the results of irradiation of a 48 cm ingot compared with a 36 cm ingot in position E2 with the flux situation of April 1982. For the 48 cm ingot, the average flux is 0.94×10^{13} n/cm^2 sec, while the 36 cm ingot receives 0.97×10^{13} n/cm^2 sec. The variation between the highest and the lowest flux is as high as 10 percent in the 48 cm case, compared to 3.8 percent for the 36 cm ingot. If the diameter is fully utilized (3 inches), the latter case represents a charge of 4.3 kg silicon.

During the period between the calibrations, maximum three months, the flux is monitored closely by means of resistivity measurements, cfr. chapter 7.

7. CONTROL OF RESISTIVITY

In the control program the actual fluxes are calculated from the resistivities of the crystals and compared with the cobalt calibration in the same height of the same channel by using the formula:

$$Q = \frac{F_{Si}}{F_{Co}} = \frac{C}{T \cdot F_{Co}} \times \left(\frac{1}{\rho_{NTD}} - \frac{1}{\rho_{start}}\right)$$

where

Q = control value

F_{Si} = flux as calculated from the silicon resistivity

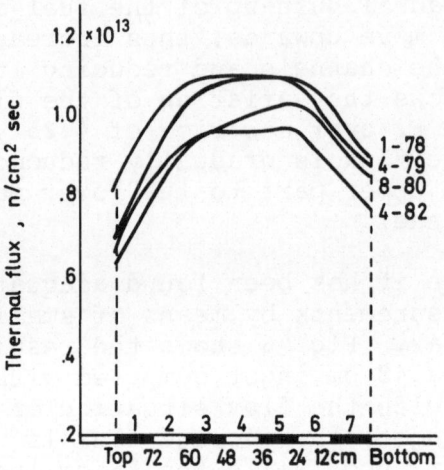

Fig. 5.   Thermal neutron flux in channel E2: January
          1978, April 1979, August 1980, and April 1982

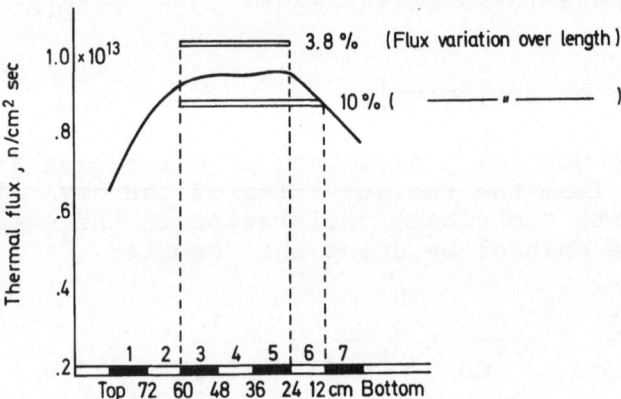

Fig. 6.   Irradiation of silicon ingots in E2, April 1982
          4 positions (each 12 cm) = 48 cm length
          3 positions (each 12 cm) = 36 cm length

F_{Co} = flux determined by means of cobalt calibration

C   = constant linking the radioactivity of cobalt with the resistivity of silicon

T   = time

ρ_{NTD} = resistivity (23°C) after irradiation

ρ_{start} = resistivity (23°C) of starting material (if n-type)

If the starting material is p-type, the resistivity is converted to an equivalent negative n-type resistivity by multiplying by the mobility ratio (- 2.89).

The resistivities for the control values are those measured at the center of the end faces on a representative number of ingots. The measurements are performed with a linear four-point probe with a probe spacing of 1.592 mm, the measurement current is 10 µA before irradiation and 100 µA after irradiation.

In order to ensure the measurement accuracy the resistivity of a standard NTD-crystal is measured every day. From this measurement it is known that the relative standard deviation on a single measurement on NTD-material is 1.1 percent. Considering the uncertainty of the irradiation time and the resistivity of the starting material, it was predicted that the control values will exhibit a standard deviation of 1.4 percent.

During the first quarter of 1982 the Topsil production irradiated in the Norwegian reactor consisted of 495 crystals, arranged in 27 groups, to obtain target resistivities between 135 and 30 ohm-cm. The size of the groups varied from 49 irradiations to 1. From this production a total of 719 control values were calculated.

The actual standard deviation on the control values was determined by comparing end faces which had been next to each other during irradiation. Eighty-five such pairs were analyzed, and from these, the relative standard deviation was calculated to be 1.8 percent, which is in good agreement with the expected value of 1.4 percent.

The overall distribution of the control values is shown in Fig. 7. They follow the bell-shaped Normal distribution quite well, with an average of 102.2 percent and a standard deviation of 4.3 percent. The irradiation

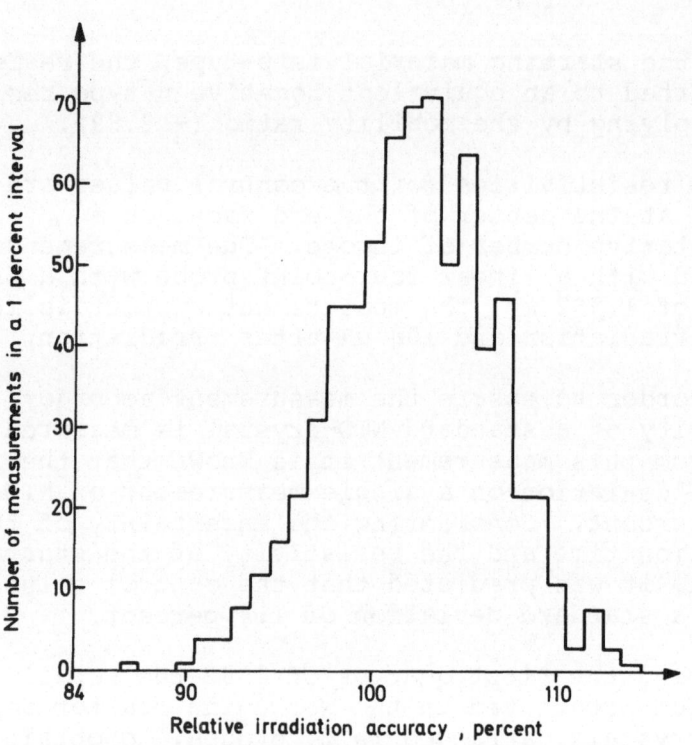

Fig. 7.   Relative irradiation accuracy in the first
           quarter of 1982
           Total number of measurements: 719
           Average: 102.2 percent
           Standard deviation: 4.3 percent

program described was comprised of irradiation lengths (single ingot or sum of ingots irradiated together) from 225 mm to 705 mm, with an average of 465 mm. Fig. 7 represents our present choice of utilization in the case of the reactor JEEP II. A lower volume utilization would have resulted in a lower standard deviation and a narrower bell-shape of the curve.

It is instructive to compare the present results with those reported by S.L. Gunn et al. at the 1978 conference (1) for the University of Missouri Reactor Facility, and by Heydorn and Andresen at the 1980 conference (2) for the Danish reactor DR 3, in which they reported a standard deviation of 3.5 percent for silicon crystals up to 400 mm length.

As mentioned above the distribution of axial flux in the irradiation channels will shift somewhat during a calibration period, because the main control rods of the reactor are slowly raised as the uranium fuel is consumed. In general, this is expected to increase the flux at the top of the channels, and reducing it in the lowest part. Because of symmetry effects, however, this pattern will not occur for all the channels. Generally this flux variation has not influenced the irradiation accuracy to any appreciable degree during the first quarter of 1982. The most significant changes during this time period were found in the channel I4, and in Fig. 8 the control values for this channel are shown as a function of the group number. In this case the group number represents true chronology except for a few minor cases. As can be seen, the trend is toward increasing dose over the whole length of the channel by 3.2 percent. The next cobalt calibration in April showed an average increase in the flux of this channel compared to January 1982 of 3.3 percent.

A similar comparison of all channels taken together has also been examined, however, it is difficult to show clear trends in the reactor behavior because of mixed load, i.e., the group number is not representing true chronology. Some of the results of the resistivity measurements indicate variations of the thermal effect of the reactor from time to ime, but such a conclusion is not supported by the reactor operation logs for the period.

Finally it has been found that the average relative dose varies only slightly from one channel to the

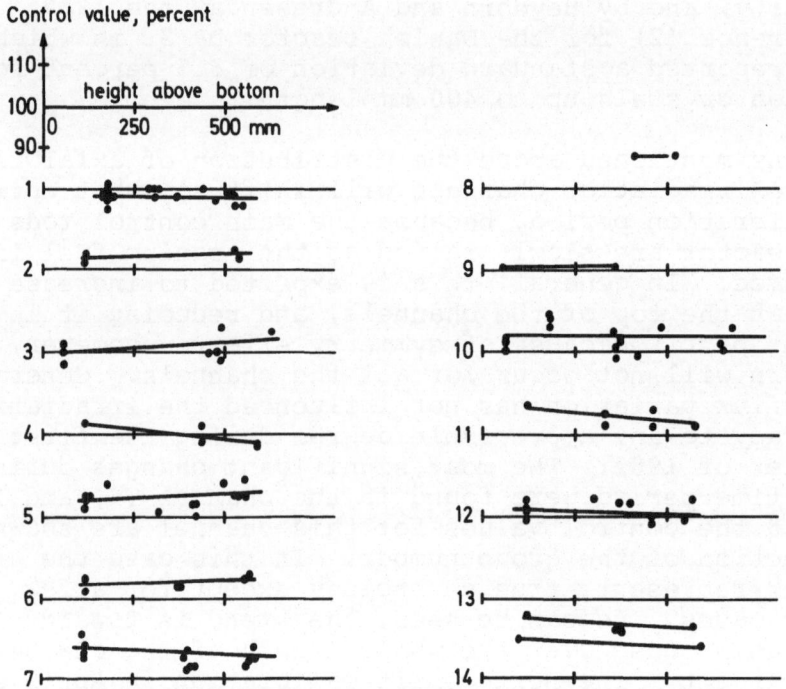

Fig. 8. Silicon irradiations in channel I4
         Axial variations of the control value during
         the period January-April 1982 (1-14 in
         chronological order)

others. No channel average deviates more than 1 percent
from the common mean value.

8. CONCLUSION

In general, it can be concluded that neutron
transmutation doping in the Norwegian reactor JEEP II
has turned out favorably. The cooperation between the
silicon producer Topsil A/S and the reactor owner
Institute for Energy Technology has now been in effect
for seven years, and is regulated by a special agreement.
The various performance data from resistivity control
measurements and reactor flux calibration will continue
to be followed closely in the future.

REFERENCES

1. S.L. Gunn, J.M. Meese, and D.M. Alger: High Precision
    Irradiation Techniques for NTD Silicon at the
    University of Missouri Research Reactor (2nd Int.
    Conference on NTD, Columbia, Missouri, USA 1978).

2. K. Heydorn and K. Andresen: Precision and Accuracy
    of NTD Silicon Production Based on Calorimetric
    Neutron Dose Control (3rd Int. Conference on NTD,
    Copenhagen, Denmark 1980).

THE DEVELOPMENT OF THE MARKET FOR NEUTRON TRANSMUTATION DOPED

SILICON

Heinz Herzer and Fritz G. Vieweg-Gutberlet

Wacker-Chemitronic GmbH
P.O. Box 1140
D-8263 Burghausen / West Germany

INTRODUCTION

As mentioned many times during this conference, neutron transmutation doped silicon was introduced to the electronic device market in the 1975-1976 time period, just seven years ago. Today, neutron transmutation doping is definitely a mature technoloy applied
mainly to semiconductor power devices. There is no doubt that the
power device sector will remain the major consumer of NTD silicon
in the future. But how about other applications and the market
growth in general?

- Do we expect the application of NTD to be spread over other
  segments of the market?
- What is the growth rate in the power sector?
- Is there enough capacity for new applications of NTD silicon?

1. MARKET DEVELOPMENT

Fig. 1. shows the dramatic rise of the consumption of NTD silicon since its introduction to the market in 1975-1976 as seen from
the point of view of Wacker-Chemitronic. It is apparent that the
1981s quantities are an order of magnitude larger than those of 1976,
when the first experimental application was for thyristors. The first
applications were related mainly to high power/high performance

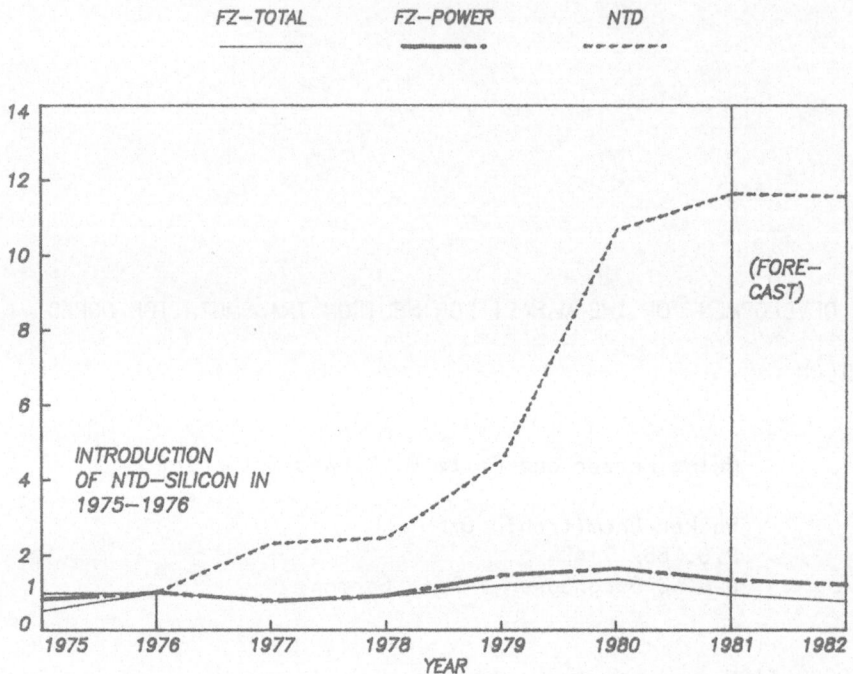

Fig. 1.   The development of float-zone silicon and NTD-silicon
          (Wacker-Chemitronic) 1975-1982 (normalized for 1976=1)

thyristors used in high-voltage direct current power transmission
systems.

      Fig. 2. shows a forecast of worldwide consumption up to the
year 1990, the end of the decade. It is the world market in real
figures (metric tons) from two points of view:

   - "BEST CASE", ending up at or around 100 metric tons per year
     at the end of the decade
   - "WORST CASE", ending up at 50-55 metric tons per year.

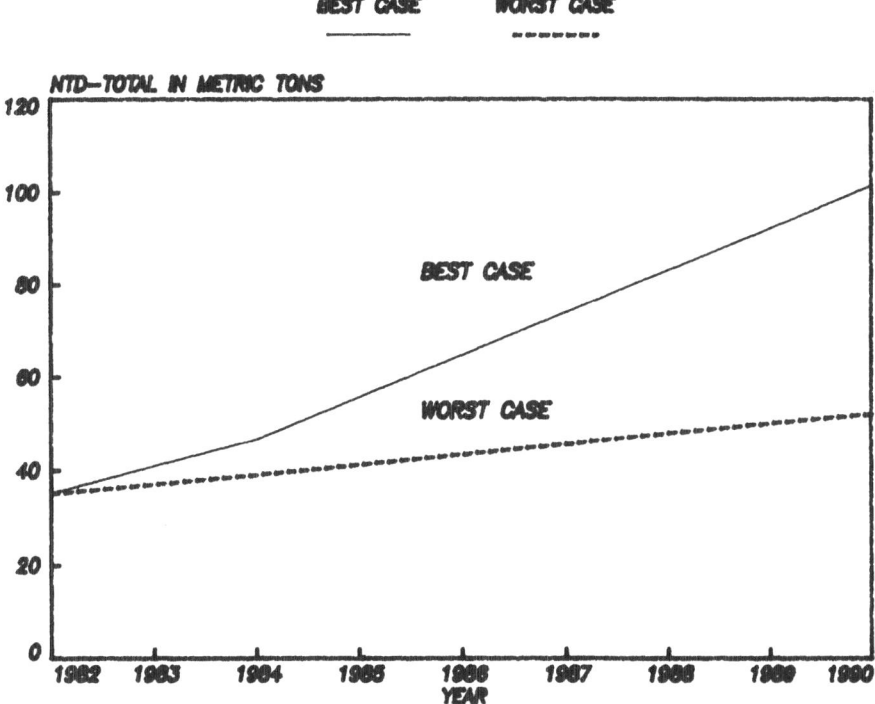

Fig. 2.   Forecast for NTD-silicon consumption worldwide 1982-1990

"BEST CASE" is based on two major assumptions:

- The power device market continues to have a steady growth rate.
- New applications for NTD silicon show up within this period
  of time.

"WORST CASE" is based on a very slow growing (even stagnant) market for power devices with no additional applications for NTD silicon in the future.

What is the real situation in the device market?

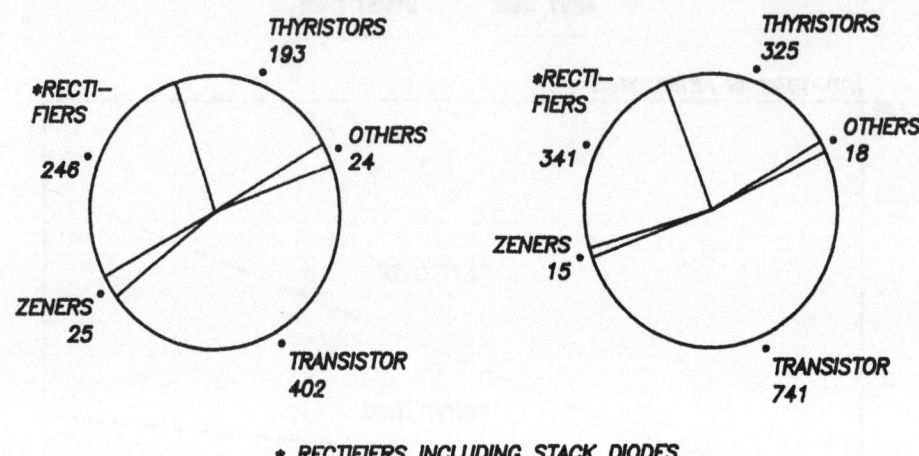

* RECTIFIERS INCLUDING STACK DIODES

1980
890 MILLION U.S. DOLLARS

1985
1440 MILLION U.S. DOLLARS

OVERALL AVERAGE GROWTH RATE ASSUMED FOR 1980–1985 10.1% p.a.

Fig. 3.   European power-semiconductor sales by product type
          (millions of US dollars) 1980-1985[1]

Fig. 3. shows a forecast for the European power device market
from 1980 to 1985[1]. The overall growth rate is expected to rise from
890 millions of US dollars in 1980 to 1440 millions of US dollars
in 1985. The fastest growing segments are transistors and thyristors
due to an extreme growth of automation and power devices needed as
the driving device for actuators.

Also with respect to future energy savings (e.g., household
equipment), power devices will be designed for the lowest energy
losses and will have tighter tolerances (e.g., with respect to base
width and base losses). These devices will require the uniformity

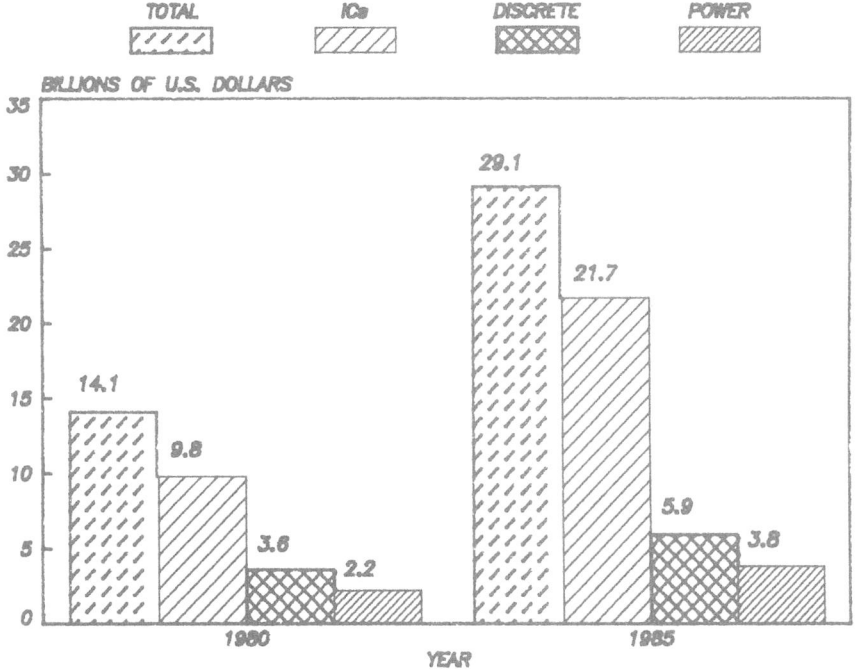

Fig. 4.   A comparison between worldwide consumption of different
          semiconductor devices 1980-1985 [2]

of resistivity that can presently be obtained by neutron transmutation doping only.

How about the market of other semiconductor devices besides
power devices?

Fig. 4. summarizes the expected growth of the worldwide semiconductor market up to the year 1985 as forecast by DATAQUEST[2] . The
quantities of discrete devices and IC's will double within this
period of time as will power devices. The question arises whether
NTD silicon will find its way into other device applications as

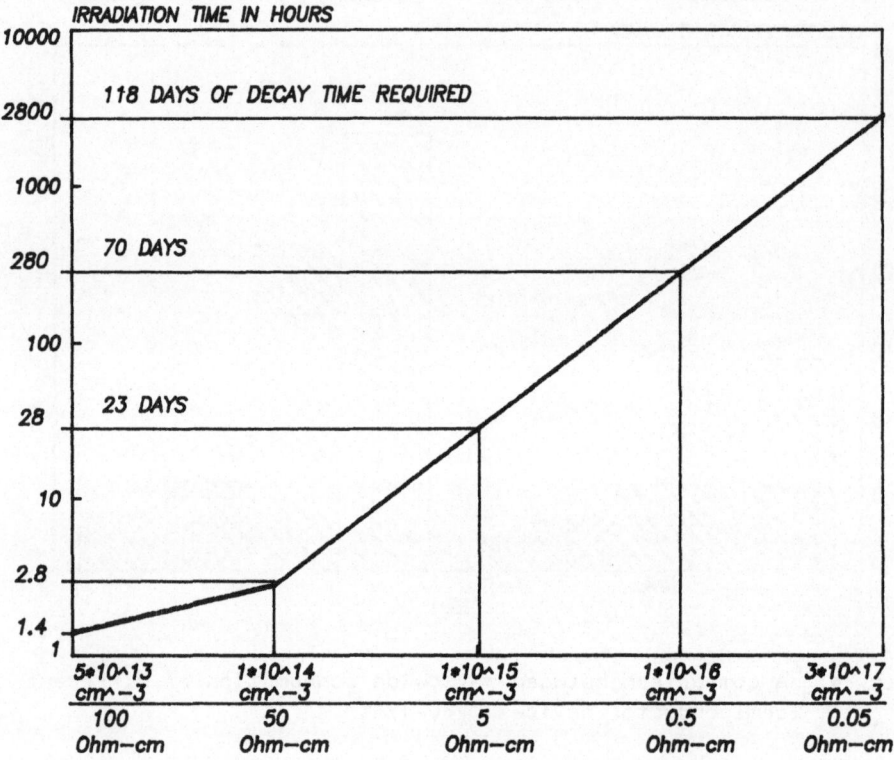

Fig. 5.   Irradiation time vs. target resistivity

power devices and, if so, what are the limitations?

Integrated circuits mainly will call for a resistivity at or around 5-10 Ohm-cm in comparison to power devices where resistivity ranges between 50 and 100 Ohm-cm (or even more).

On the other hand, new sensor devices (e.g., temperature/pressure sensors for automotive applications) will require resistivities in the same Ohm-cm range and will gain an increasing market share. So the growth rate for NTD silicon is promising for relatively low resistivity material.

Which leads us to the question: Do we have the reactor capacity for the irradiation of large quantities of silicon for relatively low target resistivities?

2. REACTOR CAPACITY

T.G.G. Smith's paper[3] presented his estimates of limited reactor capacity. Fig. 5. summarizes what lower target resistivity will mean with respect to irradiation time, decay time, and reactor capacity. When T.G.G. Smith discussed his estimates, he pointed out clearly that reactor capacity will be limited to 100 metric tons per year under the assumption that no dramatic shift in target resistivity will occur. He also mentioned that in case of a continuing development of tighter device design and tighter resistivity tolerances, other ways or other technologies besides NTD will have to be developed. However, at the present time there is no technology other than NTD known for circumventing resistivity variations occurring in crystals grown from the melt, so it is difficult to predict what these other technologies will be.

In the device manufacturing process there might be some help by more application of epitaxy or ion-implantation techniques in the future. However, as epitaxy and ion-implantation definitely are solutions to process problems in IC manufacturing, we should concentrate on power devices and sensors only.

It is not easy to understand why tighter resistivity tolerances are required in some cases of sensors (temperature/power) made of silicon. It is well known how easy it is to balance a resistivity bridge circuit by adjusting the resistor network to the resistance of the sensor. But for new applications of sensors (e.g., in automotive applications), the sensor must be a high quality product and, in case of failure, the device has to be replaced in a local repair workshop under very rough conditions. Probably no one in this workshop will be able to readjust a sophisticated resistor network in the car electronics. Therefore, one sensor (device) is of the same resistance as the other one and readjustment of the bridge circuit does not become necessary. Therefore, especially in this application, sensor devices must be made of silicon with tight resistivity tolerances - which definitely leads us to NTD silicon.

As was said before, the application of neutron transmutation doped silicon in IC manufacturing will be limited by the low resistivity and, in turn, the limited reactor capacity. Fig. 6. shows

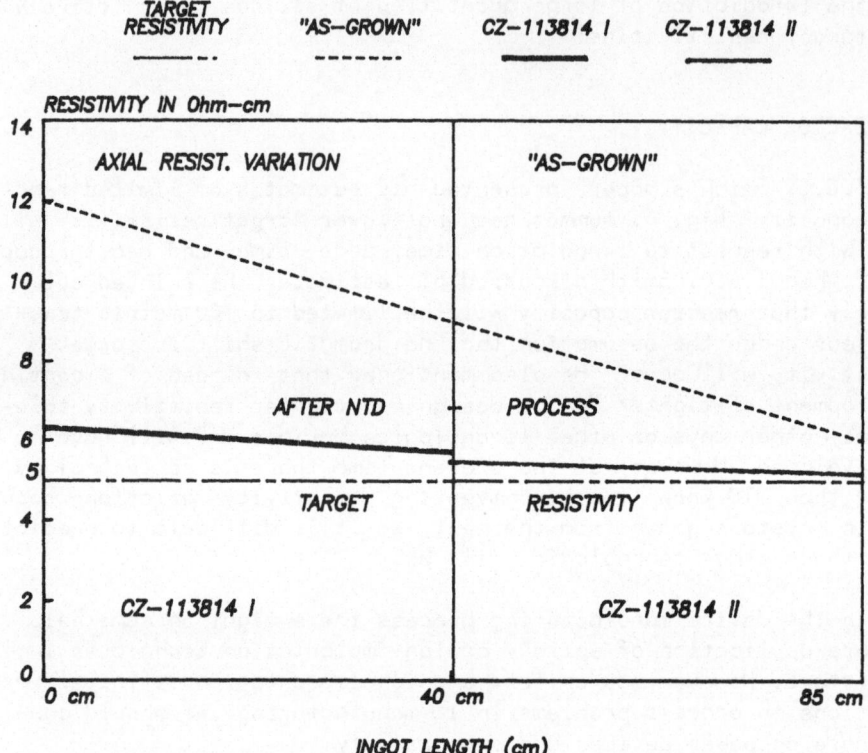

Fig. 6. Axial resistivity profile of a CZ ingot

a possible way to overcome resistivity variations - mainly axial
resistivity variations in Czochralski grown silicon ingots. As a
result of the distribution coefficient of phosphorus in silicon,
the resistivity distribution over the length of a CZ grown ingot
shows a variation with a lower resistivity at the tang end of the
ingot. This variation limits the usage of the entire ingot grown
in this process for applications with tight resistivity tolerances.
By controlled irradiation of the ingot cut into 2 (or even 3) segments, the resistivity of those segments can be adjusted to the target resistivity for a given device. This process does not call for
increased reactor capacity but will increase the yield of irradiated crystals considerably.

3. SUMMARY

The development of the silicon semiconductor device market may open a wide variety of future applications of NTD silicon.

Unless new reactor capacities will become available by the end of the decade, NTD silicon application will probably be limited mainly to power and sensor devices.

With new reactor capacities, new and promising applications may occur in the future. So the sky will be the limit! Take your choice.

REFERENCES

1   Frost & Sullivan, 'E411, The Power Semiconductor Market in Western Europe, Table 5.1
2   Dataquest, Inc., August 1981 SIS Volume III
3   T.G.G. Smith, This Volume

NEUTRON TRANSMUTATION DOPED SILICON FOR

POWER SEMICONDUCTOR DEVICES

(INVITED PAPER)

B. Jayant Baliga

General Electric Company

Schenectady, NY 12345

ABSTRACT

Power semiconductor devices operate at high currents and voltages. The fabrication of these devices requires high-resistivity silicon wafers with good spatial uniformity over large areas. Neutron transmutation doping (NTD) has been demonstrated to be an excellent technique for achieving these requirements. The development of commercially available NTD silicon has consequently led to substantial improvements in the yield of high-voltage power rectifiers and thyristors. In addition, the greater spatial uniformity, as well as the precise control over the resistivity achievable by using the NTD process, has led to a substantial increase in the breakdown voltage capability of thyristors.

This paper reviews the impact of using neutron transmutation doped silicon for power-device fabrication. Experiments conducted to verify that defects produced during NTD processing can be subsequently annealed out will be described. These experiments were crucial to allaying initial concerns regarding the quality of neutron transmutation doped silicon and were instrumental in achieving the present widespread use of this process for fabrication of starting material for the power-device industry.

The year 1982 marks the twenty-fifth anniversary of the development of the power thyristor. It is therefore appropriate at this juncture to review the impact of neutron transmutation doping (NTD) of silicon upon the fabrication of high-voltage,

167

large-area power thyristors. The basic concept of doping silicon by creating phosphorus atoms by the absorption of thermal neutrons was discussed in 1961 by Tanenbaum and Mills (1). However, it was not developed into a production process until recently. In 1975 Herman and Herzer wrote a paper (2) in which they stated that:

> "The results indicate that transmutation doping may be a rather good technique for production of homogeneously doped silicon material . . . . Further work must be undertaken to establish whether transmutation doping will play a dominant role in addition to the standard methods for growing silicon crystals."

on which Burtscher (3) commented in 1976:

> "Herman and Herzer write about the chances of neutron-irradiated silicon perhaps somewhat too cautiously . . . The use of neutron-irradiated silicon for manufacturing large area devices (especially high power devices) is by no means a question of the future . . . The use of neutron-irradiated silicon for the production of high voltage, high power devices has become standard."

Since 1976 this statement has been reinforced to the degree that almost all high-voltage, large-area power thyristors are presently fabricated by using neutron transmutation doped silicon as the starting material. This trend is expected to continue into the future.

In order to examine the impact of the development of neutron-doped silicon upon power thyristors, it is interesting to examine the growth in the power ratings of these devices over the last 25 years. The progressive increase in the blocking-voltage capability (which is related to the device breakdown voltage) and the current-handling capability (which is related to the device area) are shown in Figure 1 and Figure 2. A careful look at these figures indicates that after rather gradual increases in the blocking-voltage and current-handling capability from 1958 to 1975, these power ratings of the thyristors have shown a much more rapid increase from 1975. This upsurge in the power ratings is directly attributable to the availability of the improved starting material for thyristor fabrication as a result of the development of NTD silicon.

When the concept of using neutron-doped silicon for power devices was initially proposed in the early 1970s, it was clear that significant improvements in device power ratings would accrue

2

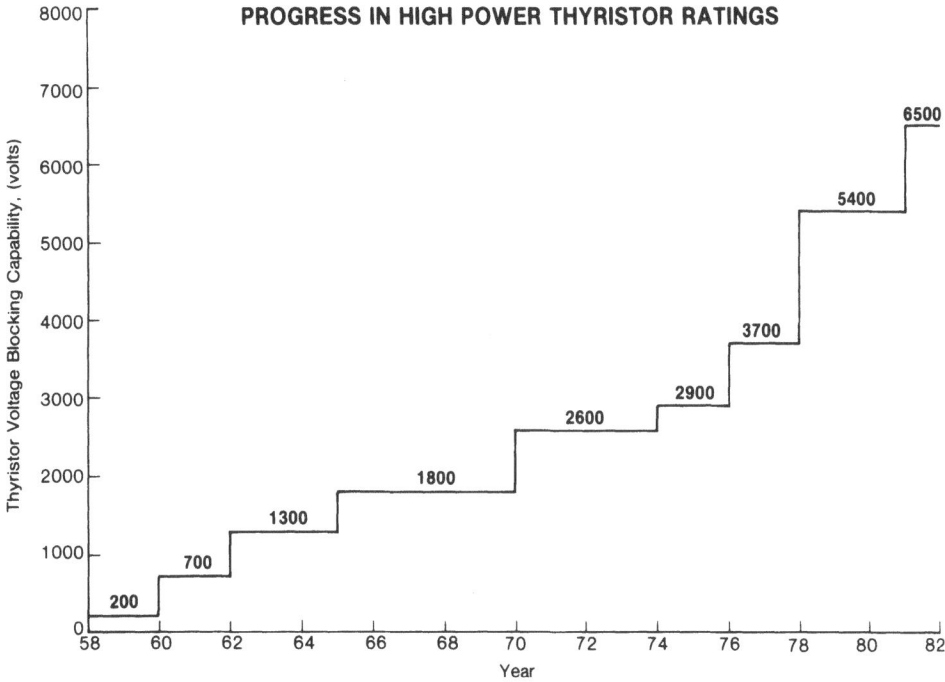

Figure 1: Progressive increase in the blocking voltage capability of power thyristors.

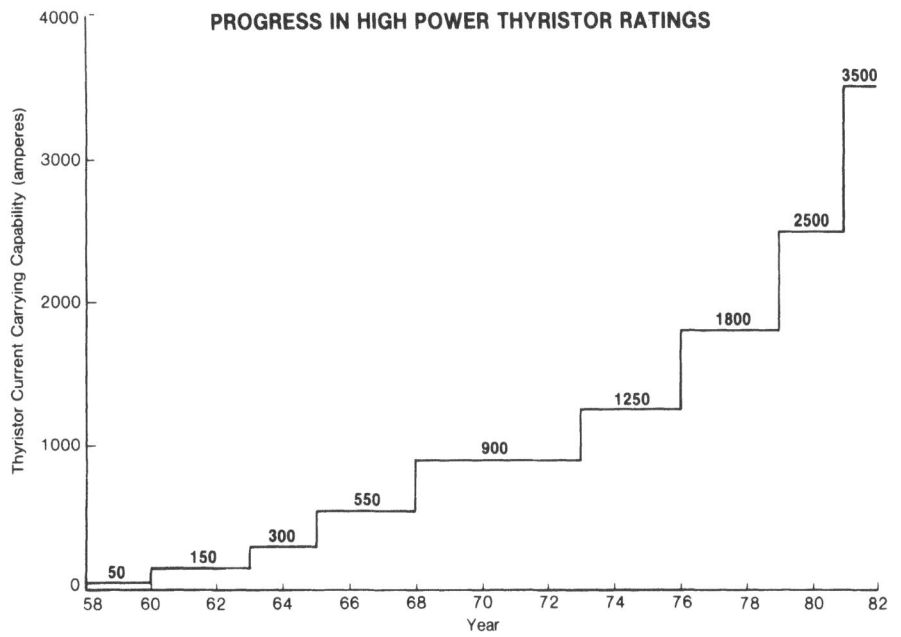

Figure 2: Progressive increase in the current-handling capability of power thyristors.

from the projected improvement in doping homogeneity within the
wafers. However, several important concerns with regard to this
new material were expressed by the power-device industry. These
concerns were related to the secondary reactions listed in Figure
3 which cause the production of sulfur due to thermal neutron
absorption by the phosphorus (P^{31}) and production of magnesium by
the fast neutron absorption by silicon (Si^{30}). These elements
even at low concentration levels were expected to reduce the
minority-carrier lifetime because they create deep lying levels in
the silicon energy gap which constitute recombination centers. In
addition, it was known that after the neutron doping had been performed, the silicon lattice contained a high level of damage due
to gamma recoil during the transmutation reaction and the irradiation of the crystal by the high energy β particles emitted during
the decay of Si^{31} to P^{31}. Further, the nuclear reactors in which
the thermal neutron capture was performed for the doping were
known to contain a background of fast neutrons. The collision of
fast neutrons with silicon atoms was known to create localized
damage clusters which could represent serious problems because
they might influence the diffusion front during device fabrication. This would severely degrade device breakdown voltage negat-
ing the sought-after advantages of the homogeneous doping.

NEUTRON TRANSMUTATION DOPING

MAIN REACTION:
$$Si^{30} (n, \gamma) Si^{31} \rightarrow P^{31} + \beta^-$$
$$2.62 \text{ hrs}$$

SECONDARY REACTIONS:

(1) $P^{31} (n, \gamma) P^{32} \rightarrow S^{32} + \beta^-$
14 days
(Radioactivity)

(2) $Si^{28} (n, 2n) Si^{27} \rightarrow Al^{27} + \beta$
(Negligible)

(3) $Si^{28} (n, \alpha) Mg^{25}$
       $Si^{29} (n, \alpha) Mg^{26}$

Figure 3:   Neutron Transmutation Doping Reactions.

4

NTD ANNEALING EXPERIMENTS:

Due to these concerns, the annealing of neutron-irradiated silicon was examined at many laboratories. The silicon material parameters that were examined included the resistivity, the minority-carrier lifetime, and the deep levels present within the silicon energy gap. The results of these studies are briefly discussed below with representative examples of data taken during these annealing studies.

Resistivity: After the neutron irradiation process it was invariably found that the silicon resistivity exceeded 10^5 ohm-cm. This was the direct consequence of the radiation damage which produced a large density of deep levels. Heat treating the wafers after irradiation was found to be effective in annealing out this damage (4-6). The results of isochronal annealing (6) of high resistivity (2000 ohm-cm) float-zoned (FZ) starting material and lower resistivity (5 ohm-cm, n-type and 7 ohm-cm, p-type) Czochralski (CZ)starting material are shown in Figure 4. It can be seen that the more heavily doped (lower resistivity) starting material required annealing in excess of 550°C to recover the final resistivity as compared with annealing temperatures above 650°C for the more lightly doped (higher resistivity) starting material. This is because the neutron damage induced deep level concentration decreases with increasing annealing temperature. It becomes comparable in magnitude to the higher background doping of lower resistivity material at a lower annealing temperature than for higher resistivity material i.e. it takes a higher annealing temperature to lower the deep level concentration to values comparable to the doping levels of 2000 ohm-cm material than for 5 ohm-cm material. These results indicate that annealing temperatures must generally exceed 700°C to ensure stabilization of the resistivity after neutron irradiation. It is worth noting here that power-device fabrication requires temperatures in excess of 1000°C. Consequently, unless the NTD silicon has been annealed to above 700°C prior to device fabrication, the starting material resistivity can be expected to undergo an undesirable change during device fabrication. This behavior was unfortunately experienced during some of the early studies on using NTD silicon for power thyristor manufacturing because some silicon vendors were not adequately annealing their crystals before measuring their resistivity and supplying them to the thyristor manufacturers. At that time, the wafer qualification procedure for NTD silicon sometimes included a resistivity stability test by the device manufacturer. With the improved understanding by the silicon suppliers of the ramifications of improper annealing of NTD silicon prior to device fabrication, this no longer appears to be a problem in the industry.

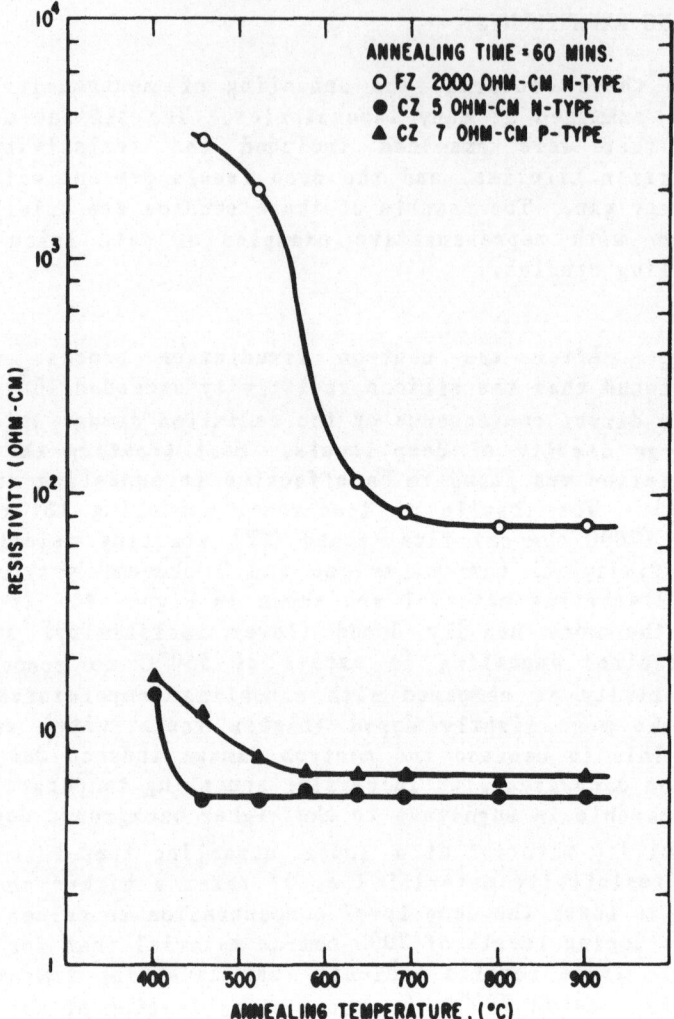

Figure 4: Variation in the resistivity of neutron irradiated FZ
           and CZ wafers during isochronal annealing between 400°C
           and 900°C (From Ref. 6).

Lifetime: As stated earlier in the paper, the minority-carrier
lifetime was another major concern for power-thyristor manufacturers. A high minority carrier lifetime is necessary in these
devices in order to obtain devices with acceptably low forward
voltage drops during current conduction. A high minority-carrier
lifetime also results in low device leakage currents in the blocking state. These device features minimize the power losses in the
devices which simplify the heat sinking of the devices, thus
decreasing the system cost.

The minority-carrier lifetime in both p- and n-type neutron-irradiated silicon was examined by using reverse recovery measurements of diodes fabricated in low-resistivity starting material prior to the neutron irradiation (6). Figure 5 shows the increase in the minority-carrier lifetime during isochronal annealing between 400°C and 900°C. The minority-carrier lifetime was observed to increase from about 0.1 microseconds after annealing at 400°C to more than 20 microseconds after annealing at above 700°C. It was found that unirradiated control samples had minority-carrier lifetimes comparable to the plateau in the lifetime data at above 700°C shown in Figure 5 for the irradiated samples.

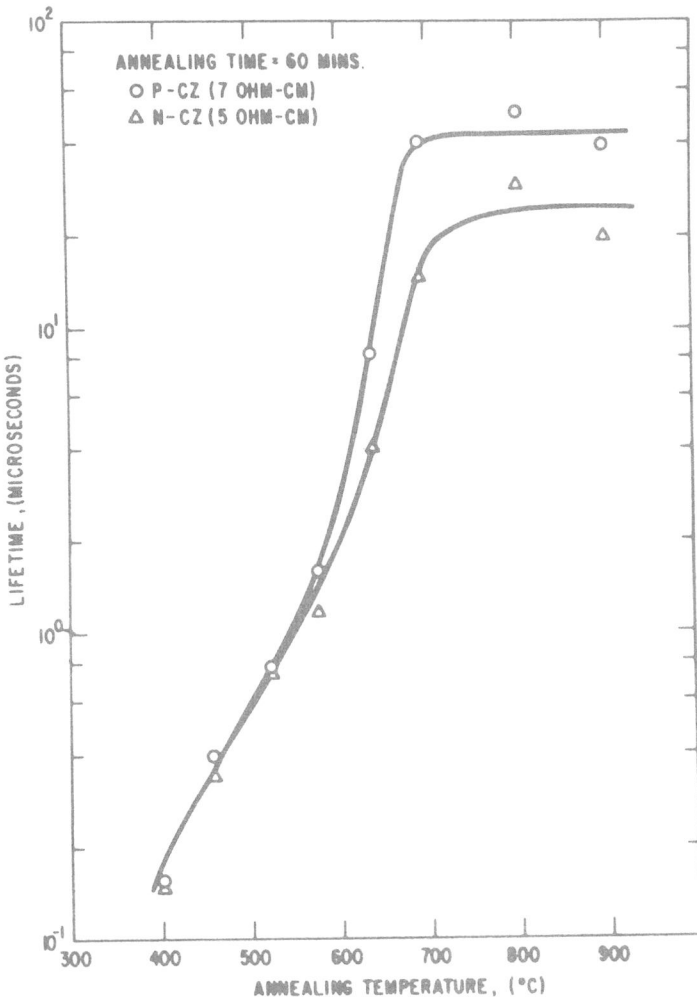

Figure 5:   Increase in the minority-carrier lifetime of neutron-irradiated silicon during isochronal annealing between 400°C and 900°C. (From Ref. 6)

Thus the lifetime in the plateau region is limited by the background contamination in the diode processing and annealing fur-
naces and is not limited by the neutron-irradiation-induced damage. These lifetime values are adequate for obtaining good power-
device characteristics making the annealed NTD material acceptable
for power-device fabrication.

The effect of fast-neutron damage upon the lifetime of silicon wafers doped to various resistivity levels has also been examined (7). Figure 6 shows the lifetimes observed after annealing
of NTD wafers irradiated in 3 different reactors. Reactor I was a
graphite/heavy-water reactor with a Cd ratio of 1000 to 2000 at
the irradiation position; reactor II was a graphite/heavy-water
reactor with a Cd ratio of 20 to 100 at the irradiation position;
and reactor III was a swimming-pool reactor with a Cd ratio of 5-
10 at the irradiation position. The data in Figure 6 indicate
higher lifetime NTD material can be produced in swimming-pool
reactors as compared with the graphite/heavy-water reactor. This
is an unexpected result because a larger Cd ratio should favor
higher lifetime values due to the smaller level of the fast neutron damage. No explanation of these results has been proposed.
However, the most important conclusion to be derived from the data
in Figure 6 is that all these methods of neutron doping produce
material with lifetimes adequate for power device fabrication.

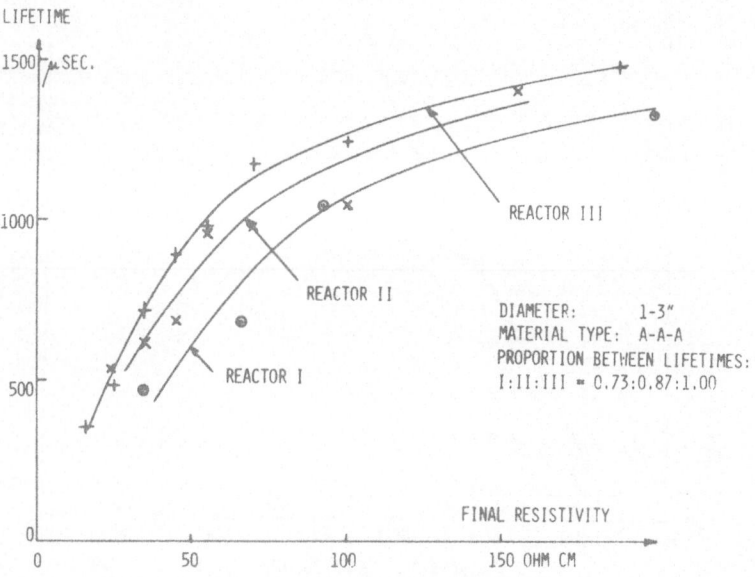

Figure 6:   Average lifetime of annealed NTD silicon produced in 3
            reactor types. (From Ref. 7)

Deep—Level Studies: In order to obtain an understanding of changes in the resistivity and the minority-carrier lifetime during the annealing of NTD silicon, it is important to examine the deep levels in the silicon band gap. These deep levels act as compensating centers which affect the resistivity and as recombination centers which affect the lifetime. Several studies of deep levels arising from neutron irradiation have been undertaken (6,8-10). The results of these studies have been summarized in Figure 7. At least four majority-carrier trap levels have been identified on each half of the silicon band gap. To obtain an indication of which of these levels is of importance in controlling the lifetime, correlation of the annealing rate of the levels with the lifetime recovery during isochronal and isothermal annealing was attempted (6). Unfortunately, all the levels were found to exhibit comparable annealing rates, making this correlation uncertain. As an alternate but less conclusive method, the dependence of the lifetime upon the ambient temperature was measured to obtain an estimate of the dominant deep-level position. It was found that the level E_v + 0.30 eV is dominant for n-type silicon and the level at E_c - 0.30 eV is dominant for p-type silicon.

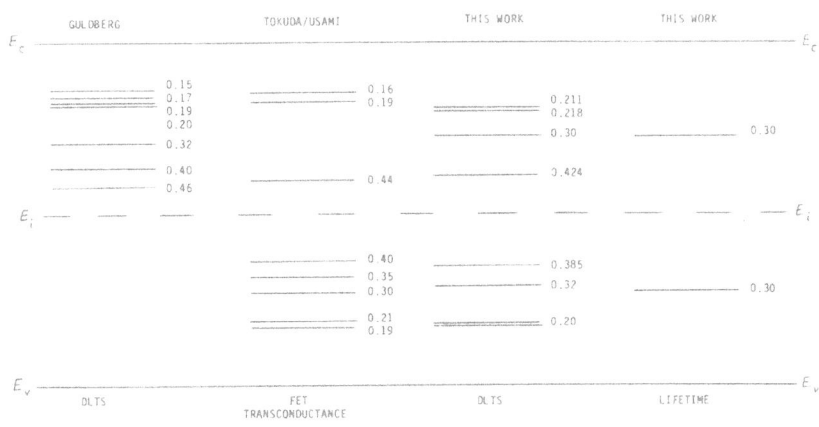

Figure 7: Comparison of deep levels observed in neutron-irradiated silicon (From Ref. 6)

IMPACT OF NTD ON POWER DEVICES

In order to understand the impact of neutron transmutation doping upon power devices, it is important to analyze the relationship between device characteristics and the starting-material properties. Typical cross sections of a power rectifier and a power thyristor are shown in Figure 8. In both of these devices, the voltage is supported during the blocking state by the formation of a depletion layer in the lightly doped n-base region whose doping (N_D) and width (W_D) are determined by the starting material.

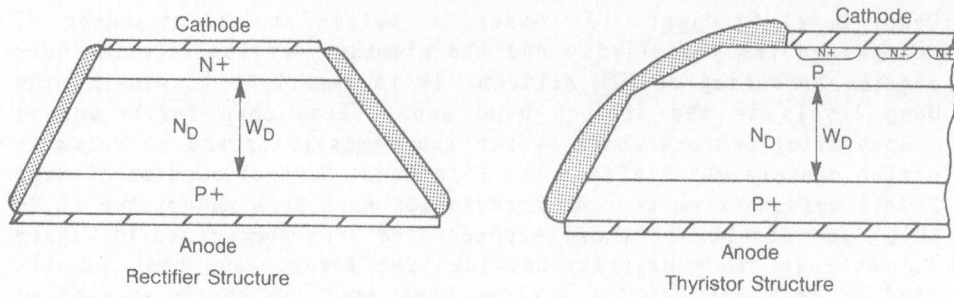

**Figure 8: Cross sections of a power rectifier and a power thyris-
tor.**

In the case of the power rectifier, the optimum device design
requires the doping level N_D to be chosen so as to obtain the
desired breakdown voltage for the device. This breakdown voltage
usually exceeds the desired operating blocking voltage by up to 25
percent and is related to the doping level by the expression (12).

$$BV_{APP} = 5.34 \times 10^{13} (N_D)^{-3/4} \qquad (1)$$

where the BV_{APP} represents the breakdown voltage of an abrupt
parallel-plane junction. In addition, the base width W_{DR} is
chosen to be equal to the depletion-layer width at the operating
voltage (V_o):

$$W_{DR} = W_C \sqrt{\frac{V_o}{BV_{APP}}} \qquad (2)$$

where

$$W_C = 2.67 \times 10^{10} (N_D)^{-7/8} \qquad (3)$$

is the depletion width at breakdown. This design for the power
rectifier allows the depletion layer to extend from the P+ anode
to the N+ cathode at the peak operating voltage. In the case of
the power thyristor, this punch-through design cannot be used
because the device contains a parasitic open-base p-n-p transistor. The breakdown voltage of this transistor is dependent upon
the n-base doping level and upon the n-base width as illustrated
in Figure 9. At the higher doping levels and for small minority-

carrier diffusion lengths (L$_p$), the breakdown voltage approaches the avalanche limit. At the lower doping concentrations, it is determined by the punch-through limit given by:

$$BV_{PT} = \frac{qN_D W_n^2}{2\varepsilon\varepsilon_o}.$$

(4)

It can be seen from Figure 9 that there is an optimum doping level at which a peak in the breakdown voltage is observed. If this peak in the breakdown voltage is achieved, the n-base width will also be minimized. This impacts both the forward voltage drop during current conduction and the turn-off time of the rectifier and the thyristor as illustrated in Figure 10. This figure shows trade-off curves between the forward voltage drop and the turn-off time for devices of various n-base widths. Since it is desirable to simultaneously decrease both the forward drop and the turn-off time in order to minimize the power dissipation, it is clear that a narrower n-base width is highly desirable.

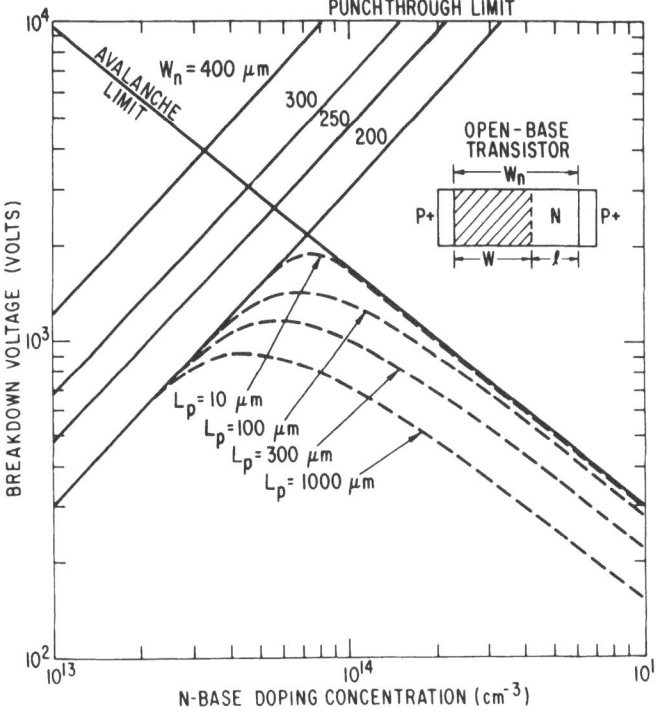

Figure 9: Open-base transistor breakdown voltage as a function of the doping of the n-base and its width. W is the depletion width.

11

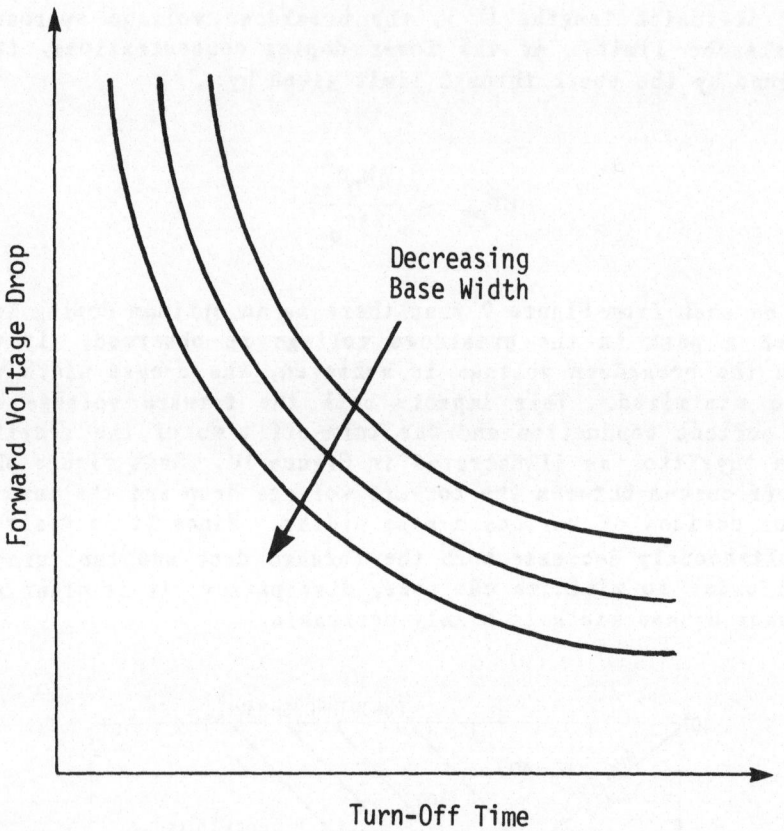

Figure 10: Trade-off curves between forward voltage drop and turn-off time for various n-base widths.

A comparison between the resistivity variation in conventionally doped silicon and NTD silicon is shown in Figure 11. Conventionally doped silicon showing local doping variations of ± 15 to 20 percent compared with the less than ± 2% typical of NTD silicon. The consequence of this is that in NTD silicon the device designer can obtain the peak in the breakdown voltage indicated in Figure 9. By contrast for conventionally doped silicon, allowance must be made for both the lowest and highest resistivity in the slice. The lowest allowable resistivity must be chosen to exceed the avalanche breakdown limit while the n-base width must be chosen to exceed the depletion width at the highest resistivity point in the slice. This compromise results in achieving higher breakdown voltages for the same n-base width for NTD wafers compared with conventionally doped wafers. These devices can, there-

fore, handle higher voltages for the same forward drop and turn-
off speed.   Chu and Johnson (11) have reported an increase in
thyristor breakdown voltage from 3000 volts to 3500 volts for the
same forward drop by converting from conventionally doped silicon
to NTD silicon.   They have also observed a decrease in the stored
charge during reverse recovery.

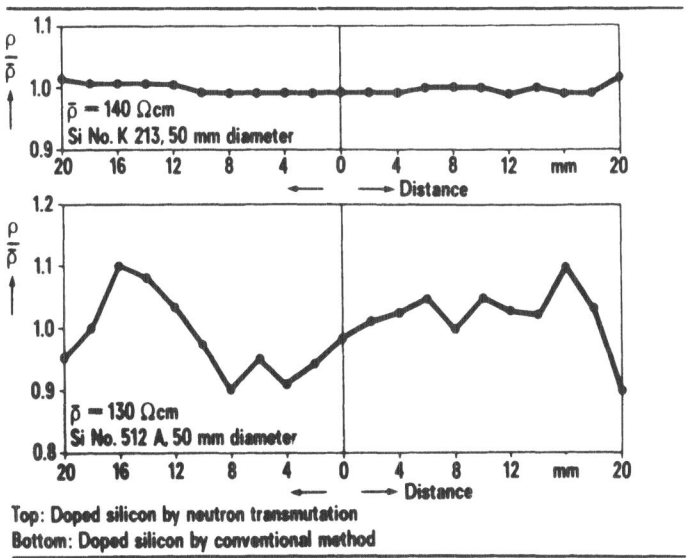

Figure 11: Typical resistivity fluctuations for conventionally
           doped and NTD silicon (From Ref. 11)

CONCLUSIONS

Neutron transmutation doped silicon became a commercially
available product in 1976. It was immediately recognized that the
improved resistivity homogeneity achievable in this process should
greatly benefit power-thyristor characteristics and their production yields. The major concerns in the power industry before
accepting this material were related to the neutron-irradiation
damage, the residual radioactivity, and the minority-carrier lifetime in the starting material. It has now been conclusively
demonstrated that these potential problems can be overcome by adequate thermal annealing procedures prior to device fabrication.
Neutron transmutation doped silicon has therefore been well
accepted as a superior starting material compared with conventionally doped silicon for the manufacturing of power devices. Its

13

introduction has not only led to improvements in the ratings and yields of the devices which were being manufactured in 1976 but has significantly accelerated the development of power devices with higher voltage blocking capability and current handling capability. Since no breakthroughs in improving the homogeneity of conventionally doped crystals are anticipated, it is expected that neutron transmutation doped silicon will continue to be used by power device industry in the foreseeable future.

REFERENCES

[1] M. Tanenbaum and A.D. Mills, J. Electrochem. Soc., Vol 108 , pp.171-176 (1961).

[2] H.A. Herman and H. Herzer, J. Electrochem. Soc., Vol 122 , pp.1568-1569 (1975).

[3] J. Burtscher, J. Electrochem. Soc., Vol 123 , pp.948-949 (1976).

[4] H. Herzer, in 'Semiconductor Silicon,' pp.106-114 (1977).

[5] F.A. Selim, P.D. Blais, P. Rai-Choudhury, and R.F. Yut, in 'Semiconductor Silicon', pp.126-134, (1977).

[6] B.J. Baliga and A.O. Evwaraye, in 'Neutron Transmutation Doping in Semiconductors,' pp.317-328, Ed. J.M. Meese, Plenum Press (1979).

[7] O. Malmros, in 'Neutron Transmutation Doping in Semiconductors', pp.249-259, Ed. J.M. Meese, Plenum Press (1979).

[8] J. Guldberg, Appl. Phys. Lett. Vol 31 , pp.578-579 (1977).

[9] Y. Tokuda and A. Usami, J. Appl. Phys., Vol 47 , p.4952 (1976).

[10] W.E. Haas and M.S. Schnoller, J. Electronic Materials, Vol 5 , pp.57-68 (1976).

[11] C.K. Chu and J.E. Johnson, in 'Neutron Transmutation Doping in Semiconductors,' pp.53-63, Ed. J.M. Meese, Plenum Press (1979).

[12] B.J. Baliga in 'Silicon Integrated Circuits,' Editor, D. Kahng, Academic Press (1981).

PROCESS INDUCED RECOMBINATION CENTRES IN NEUTRON TRANSMUTATION

DOPED SILICON AND THEIR INFLUENCE ON HIGH-VOLTAGE DIRECT-CURRENT

THYRISTORS

D.E. Crees and P.D. Taylor

Marconi Electronic Devices Limited
Lincoln
England

ABSTRACT

The tight control of doping level and uniformity offered by
NTD Silicon has permitted the more optimised design of power
semiconductor devices. In particular large diameter high
resistivity NTD Silicon has permitted the design of higher voltage
and higher current thyristors for High-Voltage Direct-Current
applications. In this paper we report on a study of the effects
of lifetime degradation in NTD Silicon during device processing
and its influence on forward-voltage drop. It is found that the
effects of lifetime degradation in the p-base region of the
thyristor are greater than in the wide undiffused n-base region.

INTRODUCTION

NTD Silicon has now found wide acceptance as the most optimum
material for the production of high power semiconductor devices.
In particular it has allowed the device designer to minimise base
widths to satisfy more easily the competing demands of low leakage,
low on-state voltage drop and fast switching times.

In particular, in the field of High Voltage Direct Current
(HVDC) transmission applications the introduction of NTD Silicon
has made an especially strong impact. This is because HVDC
systems demand the highest voltage, highest current thyristors,
while placing severe constraints on the stored charge and turn-off
time requirements of the device. Also, there are strong incentives
to reduce operating costs by minimising power lost in the thyristor

181

valve, and for the thyristor designer this requires a tightly controlled and low on-state voltage drop. There are, therefore, significant benefits in fully utilising the resistivity homogeneity of NTD Silicon for optimisation of device design.

Partly as a result of optimising the device design, and partly due to the push by HVDC requirements to larger silicon diameters and higher blocking voltages, silicon material and processing technologies have become increasingly more critical areas of thyristor manufacture. In terms of process control the most dominating requirements for the HVDC devices are forward voltage drop, stored charge and turn-off time, and it is achieving the correct overall combination of these parameters which is the most demanding factor. In this paper we will discuss the results of an investigation to determine the main factors controlling the forward voltage drop of an HVDC thyristor based on NTD Silicon. It will be shown that the critical area of control lies in the diffused p-base lifetime in contrast to the wide undiffused n-base for which adequate lifetimes are more readily achieved.

THE HVDC THYRISTOR

The HVDC thyristor used during this study was a 3200V 1000A device based on 58 mm diameter 180 ohm-cm NTD Silicon. The outline specification for this device is given in Table 1.

Table 1. Basic HVDC Specification

Repetitive Forward Blocking Voltage	3200V at 50mA
Repetitive Reverse Blocking Voltage	4000V at 150mA
Average Current	1000A
Surge Current	17kA
dV/dt	2000V/μs 120°C
Trigger current/voltage	150mA/1.1V typ.
dI/dt repetitive	200A/μs 90°C
Reverse avalanche	4.5kV 60A peak
On-state voltage drop	1.7V 1200A
Turn off time	150μs
Stored charge	380μC typ.

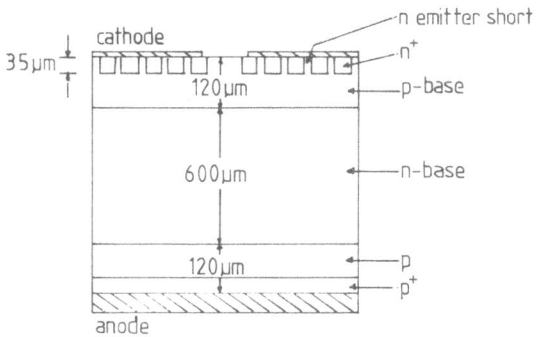

Fig. 1   Schematic cross section of the 3200V,
         1000A HVDC Thyristor

The device is shown schematically in Figure 1.

   The deep p-type diffusions are produced by a sealed tube
gallium and aluminium diffusion.   This is followed by the n+
emitter diffusion using an open-tube phosphorus system.   As the
final stage of the diffusion process the silicon is given a POCL$_3$
gettering treatment to remove metallic contaminants.   A typical
spreading resistance profile for the completed thyristor is shown
in Figure 2.

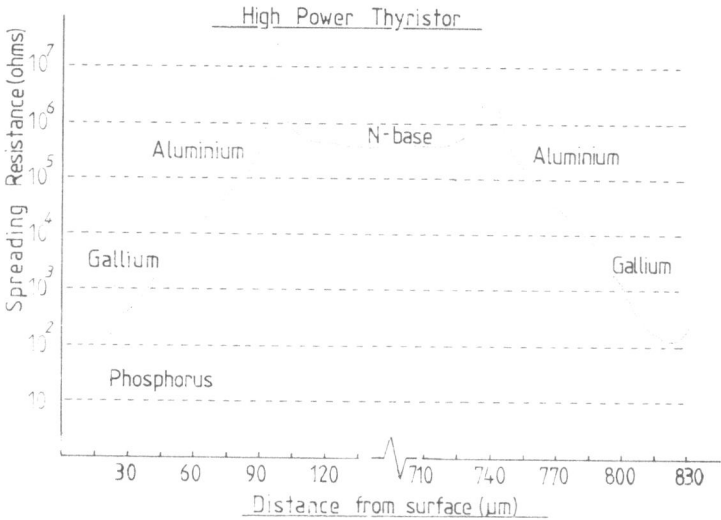

Fig. 2:   Typical spreading resistance plot for the
          HVDC thyristor

FACTORS AFFECTING THE FORWARD-VOLTAGE DROP

In high power thyristors there are three main factors controlling the forward-voltage drop: these are the base lifetime, the emitter efficiency and the turn-on uniformity.

The base lifetime influences the forward-voltage drop through the effects of conductivity modulation. A low lifetime in the base reduces the excess carrier concentration through carrier recombination, which results in less conductivity modulation therefore increasing the voltage drop across the base region. The emitter efficiency also controls the excess carrier concentration in the base: a reduction in the injected charge level can be caused for example by increased recombination in the emitter regions. The turn-on uniformity is also a critical factor and is related to the diffusion profile and the lifetime homogeneity.

All these three factors are of equal importance in determining the forward voltage drop in the thyristor, and in this paper we will report on the lifetime aspects of the problem excluding the influence of diffusion profiles, and geometric considerations such as the n+ emitter shorting pattern (see Fig. 1).

N-BASE LIFETIME

Lifetime and recombination in the n-base of the device have been assessed using open circuit voltage decay (OCVD) to measure the lifetime, and deep-level transient spectroscopy (DLTS) to determine the deep-level parameters. Both of these analytical techniques have been applied at various stages of processing of the HVDC thyristor and results collected over a twelve month period.

Typical OCVD and DLTS results are shown in Table 2 for NTD silicon after the first deep p-type diffusion and after the final POCL$_3$ gettering diffusion.

The measured emission rates have been fitted to an expression of the form:

$$e = AT^2 \exp (E_a/kT)$$

e = emission rate
A = a constant
T = temperature
E_a = the deep level activation energy
k = Boltzmann's constant

Table 2.   DLTS and OCVD Lifetime Results

Deep Level	Activation energy (eV)	A $(S^{-1} K^{-2})$	Concentration cm^{-3} After first diffusion	After final diffusion
A1	0.54	1×10^7	2×10^{11}	$< 1 \times 10^{10}$
A2	0.48	5.5×10^6	2×10^{11}	Not Detected
C	0.31	5.7×10^5	4×10^{11}	3×10^{10}
D	0.31	1.8×10^7	1×10^{12}	5×10^{10}
E	0.26	2.6×10^8	3×10^{11}	Not Detected
OCVD Lifetime Results			$5 - 10$ μs	$\geqslant 250$ μs

These DLTS measurements showed that after the first diffusion five major deep levels were present in the upper half of the band gap of the silicon at concentrations of up to typically 10^{12} cm^{-3}: for convenience we have labelled these levels A1, A2, C, D and E. After the final gettering treatment only three deep levels normally remain, A2 and E having been completely removed. The remaining deep levels (A1, C and D) were present at typically 1 to 5×10^{10} cm^{-3}.

The OCVD lifetime results support the DLTS findings: after the initial diffusion the lifetime is low, but after the gettering treatment the lifetime was found to be greater than 250 μs, an acceptable value for a thyristor of this type.

For one particular diffusion run the batch was sampled after each main high temperature treatment, the results are shown in Table 3.

After the phosphorus drive process, it can be seen that the concentrations of the deep levels have been reduced over those prior to this diffusion; also the POCL$_3$ gettering treatment was shown to be effective in removing the majority of the residual deep levels.

Although the concentrations of the residual deep levels are low, and the OCVD lifetime high, confirmation was required that these residual levels were not significantly affecting the on-state voltage drop of the thyristor. Accordingly several thyristors of known forward-voltage drop were selected for a DLTS examination. The results are presented in Figure 3 as a graph of total residual deep-level concentration versus the device forward voltage drop.

Table 3   Deep-Level Concentration (cm^{-3}) after each
           Diffusion Treatment

Treatment	A1	A2	C	D	E
Ga/Al	1.1×10^{11}	ND	1.8×10^{11}	5.8×10^{11}	5.5×10^{10}
P Deposition	ND	7.8×10^{10}	2.3×10^{11}	1.7×10^{12}	5.5×10^{10}
P Drive	5.1×10^{10}	ND	5.1×10^{10}	1.4×10^{10}	1.2×10^{10}
$POCL_3$ Getter	5.5×10^{9}	ND	9.8×10^{9}	2.1×10^{10}	ND

ND = Not Detected

It is clear that there is very little correlation between these
parameters. This suggests that the residual n-base recombination
centres are not dominant in controlling the forward-voltage drop
of the thyristor.

Comparisons have also been made between sample deep-level
concentrations for diffusion batches after the first Ga/Al
diffusion and the resultant on-state voltage drop of the thyristors

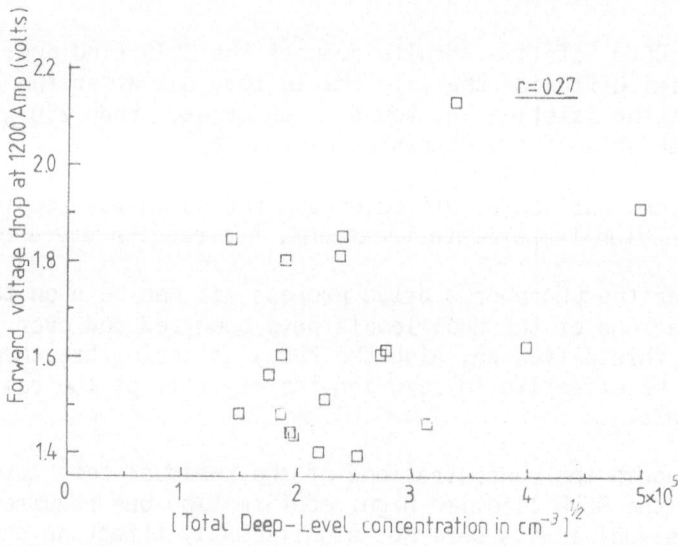

Fig. 3   Thyristor forward voltage drop as a function of the
          square root of the Deep Level trap concentration
          r = correlation coefficient.

from those batches. Out of the five observed deep levels only
level C showed a significant correlation. Figure 4 is a graph of
forward voltage drop yield for a diffusion batch as a function of
the measured concentration of level C. This result represents
an anomaly since it suggests that the forward-voltage drop yield
improves as the silicon deep level concentration is increased.
One possible explanation for this is that this particular deep level
can assist in gettering lifetime killing impurities from the silicon
during the subsequent phosphorus gettering.

Therefore from our examination of the n-base of the HVDC
thyristor we have found very little evidence to suggest that the
forward-voltage drop of the device is adversely affected by the
carrier lifetime in this region of the device.

P-BASE LIFETIME

The lifetime in the p-base of a thyristor may be conveniently
measured by applying an OCVD measurement to the n+-emitter to
p-base junction. This junction can be accessed using the gate
and cathode terminals of the device. Over a period of time we
have measured this OCVD lifetime between gate and cathode and
compared it with the device forward-voltage drop. We have found
that for a given diffusion batch a very strong correlation exists
between these two parameters and that the slice to slice variation

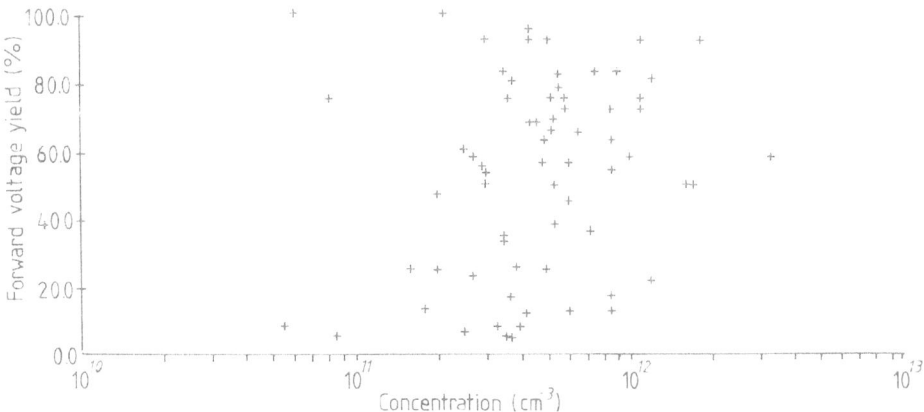

Fig. 4 Concentration of the deep level 'C' after the
        Ga/Al diffusion versus the thyristor forward-
        voltage drop yield

in forward-voltage drop is to a large extent determined by the
p-base lifetime. An example is given in Figure 5.

We conclude that, for the HVDC thyristor we have examined,
any degradation of forward-voltage drop is associated most strongly
with lifetime effect in the diffused p-base of the device. This
indicates that either the impurities responsible for the lifetime
reduction are slow diffusing species with diffusion coefficients
comparable with those of the p-type dopants gallium and aluminium
used to produce the p-base, or, that they are defect impurity
complexes incorporating gallium or aluminium. It was therefore
of interest to establish the sources of the dominant recombination
centre.

SOURCES OF CONTAMINATION

The starting NTD silicon was critically examined using DLTS
and infrared absorption techniques. A specimen DLTS plot for the
NTD starting material is shown in Figure 6. This figure shows a
broad band of defect levels at about -40°C which is attributed to

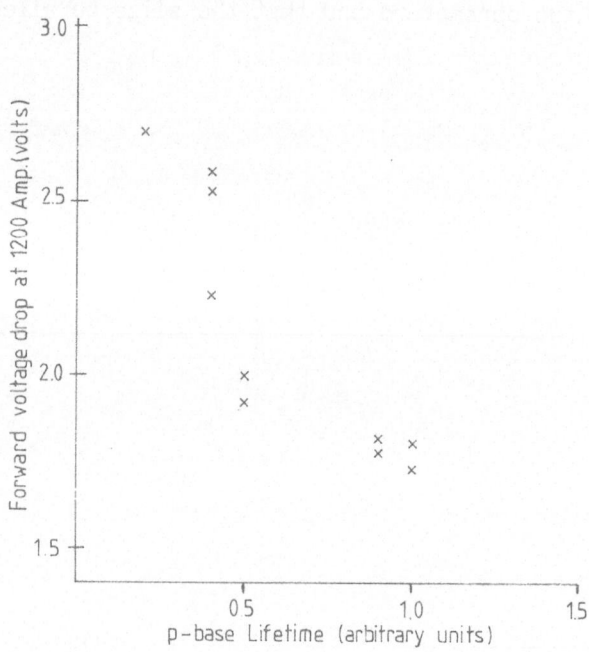

Fig. 5   Forward-voltage drop as a function of emitter-base
         lifetime for a diffusion batch

the residual irradiation damage which has not been completely
removed by the post irradiation anneal. After the first high
temperature diffusion treatment these deep levels disappear.
However, how this residual irradiation damage interacts with any
impurities diffusing into the silicon during the first Ga/Al
diffusion is an unknown factor and one which may merit further
investigation. The origin of the peak at about -170°C in Figure 6
is unknown.

To establish if any of the observed deep levels in the
silicon following the first diffusion could be affected by a
quenching treatment, a sample was quenched from 1000°C and its
DLTS properties were measured. The resultant plots taken after
various annealing times are shown in Figure 7. It was found that
only one deep level was affected by the quench and that was level
A2, which slowly decayed over a period of days.

A comparison of these results with those of Graff and Pieper
(1981) suggests that this level is associated with interstitial
iron in p-type silicon. The relevance of iron as a lifetime
killing agent in power devices is open to question because as an
element it is readily gettered out of the silicon unless allowed

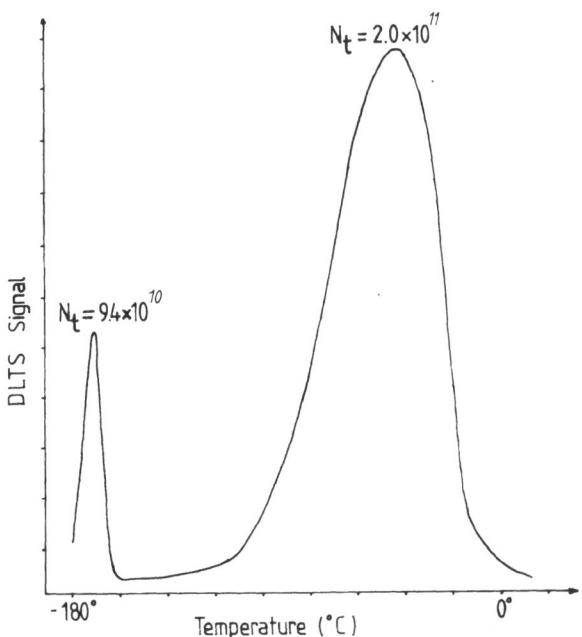

Fig. 6   DLTS plot of a sample of the NTD Silicon
         before diffusion. N_t = trap concentration
         Emission rate = 230 sec.$^{-1}$

to precipitate (e.g. by quenching). However, very little is
understood of the possible interactions between iron, silicon
and impurities, and with the presently available information it
must be regarded as an undesirable contaminant.

Level A1 is believed to be the gold acceptor level widely
quoted in literature. We have confirmed this for our thyristor
samples by comparing DLTS results from silicon before and after
a gold diffusion treatment. Like iron, gold is a very efficient
recombination level but is also a fast diffusing element and is
readily gettered from silicon.

Infrared absorption measurements of carbon and oxygen
concentration have also been made in both the silicon as received
from the supplier and in the diffused material at various stages
of process. The results in Table 4 were obtained using a Nicolet
Fourier-Transform infrared spectrometer.

These results show a steady increase in oxygen concentration
throughout the process and no carbon was detected above the
detection limit of 1.5×10^{16} cm^{-3}. However, we have found no

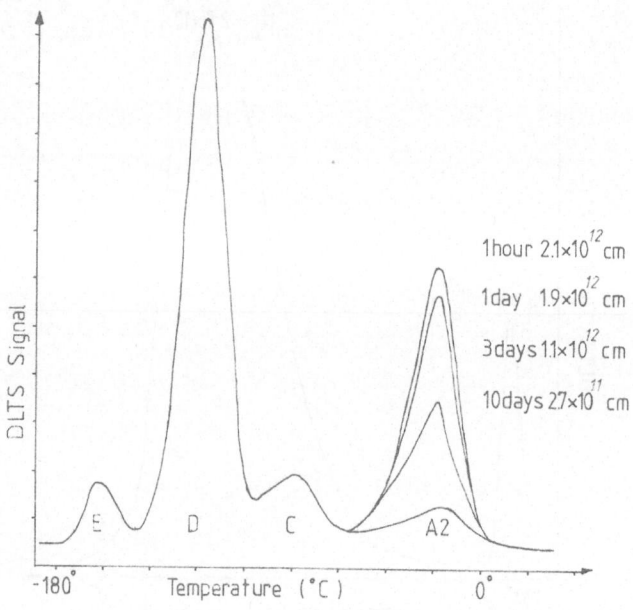

Fig. 7 Effect of annealing time at 1000°C on Ga/Al
diffused silicon. Emission rate = 230 sec^{-1}

Table 4   Infrared Absorption Measurements of Carbon and
          Oxygen Concentrations after each main process
          Stage

	Oxygen (cm^{-3})	Carbon (cm^{-3})
Incoming Silicon	$< 1.0 \times 10^{16}$	1.5×10^{16}
Ga/Al Diffused	3.8×10^{16}	"
Phosphorus Deposition	4.0×10^{16}	"
Phosphorus Drive	1.1×10^{17}	"
Getter	1.34×10^{17}	"

evidence to correlate the oxygen level to either the DLTS results
or to the measured device forward voltage drops, suggesting that
for this device the oxygen level is low enough to have little
influence on the device lifetime.

Currently, mass spectrometry and atomic absorption techniques
are also being applied to the silicon both before and after
chemical and diffusion treatments to identify the presence of
other impurities. To date, this investigation has shown iron to
be the main contaminant.

CONCLUSIONS

We have investigated the effects of lifetime on thyristor
forward voltage drop for an HVDC thyristor based on NTD Silicon.
The results of this study have demonstrated the importance of the
lifetime in the diffused p-base of the thyristor in controlling
the forward voltage drop. It is postulated that slow diffusing
contaminants or defects associated with the p-type dopants
gallium or aluminium are responsible for this lifetime degradation.
Possible sources of contamination are currently being sought and
our results to date have shown the presence of oxygen, iron and
gold in the silicon at various stages of diffusion, but have not
demonstrated that they have any effect on the lifetime in fully-
fabricated thyristors.

REFERENCES

Graff K and Pieper H, The Properties of Iron in Silicon,
J. Electrochem Soc. 128:669 (1981)

A STUDY OF FLOAT-ZONED NTD SILICON GROWN IN A HYDROGEN AMBIENT

Lu Chuengang and Li Yaoxin
Institute of Atomic Energy
Beijing, People's Republic of China

Sun Chengtai
Institute of Rectifier
Xian, People's Republic of China

Yin Janhua
Oemei Semiconductor Material Manufactory
Sichuan, People's Republic of China

INTRODUCTION

The techniques of NTD silicon production have been well developed especially for float-zoned silicon grown in an argon ambient. Irradiation effects, annealing procedures, and device fabrication techniques are well known. However, for float-zoned silicon grown in a hydrogen ambient, hydrogen is incorporated into the crystal, and it is not clear whether the presence of the hydrogen will affect the properties of NTD silicon and influence the associated device production. Experiments were performed in order to study and resolve these questions. Samples of float-zoned silicon prepared in a hydrogen atmosphere with high resistivities were selected as starting materials and were irradiated in a swimming pool reactor. Some properties of these silicon samples after this NTD doping were studied, and high power devices were fabricated and compared to similar devices fabricated from float-zoned silicon grown in an argon ambient.

INITIAL SILICON CRYSTAL

The initial silicon crystals were supplied by Oemei Semiconductor Material Manufactory. These crystals were grown by purifying the polycrystal starting material by float-zone techniques in

193

high vacuum, and then grown into a single crystal in a high purity
hydrogen ambient. The initial silicon crystals used in this study
had the following properties:

Resistivity: n-type ρ_n ⩾ 1000 ohm-cm
               p-type ρ_p ⩾ 3000 ohm-cm

Nonuniformity of the resistivities of silicon ingots along axial
and radial directions $\left[(\Delta\rho = (\rho_{max} - \rho_{min})/\rho_{max} \right]$:

  n-type $\Delta\rho_n$ ⩽ 50%
         $\Delta\rho_n$ ⩽ $(\rho_n/1000)$ x 50% (for ρ_n ⩾ 2000 ohm-cm)
  p-type $\Delta\rho_p$ ⩽ 50%
         $\Delta\rho_p$ ⩽ $(\rho_p/3000)$ x 50% (for ρ_p ⩾ 6000 ohm-cm)

Lifetime of minority carriers: ⩾ 500 μs
Crystal quality: dislocation-free and swirl-free
Orientation: <111>
Diameter: 50-60 mm
Oxygen content: <5 x 10^{16} atoms/cm^3
Carbon content: <5 x 10^{16} atoms/cm^3
Other impurities: determined by neutron activation analysis and
controlled where possible.

IRRADIATION CONDITIONS

     The initial silicon ingots were irradiated in the Swimming
Pool Reactor at the Institute of Atomic Energy. The irradiation
conditions are listed below:

Thermal flux ϕ_{th}: 3 x 10^{13} n/cm^2sec
Ratio ϕ_{th}/ϕ_f (ϕ_f--fast flux E>1 MeV): 19
Irradiation environment: H_2O
Irradiation temperature of silicon ingot: 40°C

EXPERIMENTS WITH NTD FLOAT-ZONED SILICON GROWN
IN A HYDROGEN AMBIENT

     In this study, some electrical, optical, and mechanical prop-
erties of NTD H_2-FZ Si were tested and analyzed and the results
were compared with the results of similar tests with NTD Ar-FZ Si.

Relationship Between Resistivity and Annealing Temperature

     Figure 1 shows the experimental results of measuring the
resistivities of silicon samples as a function of annealing temperature after irradiation. For H_2-FZ Si, p-n conversion occurs
at an annealing temperature of about 350°C. After this conver-

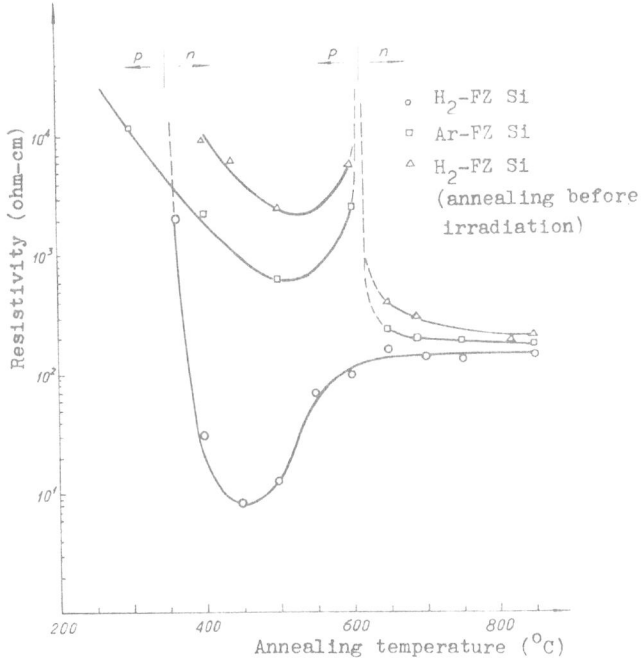

Fig. 1. Resistivity versus annealing temperature for Float-Zoned NTD Si samples grown in hydrogen and argon ambients (0.5 h iso-chronal annealing).

sion, when the temperature reaches about 450°C, the resistivity passes through a minimum which is of the order of one-tenth of the target resistivity. The resistivity then increases as the annealing temperature increases and stablizes at the target value. However, in the case of NTD Ar-FZ Si, the p-n conversion temperature is about 650°C and the minimum in resistivity does not occur.

The phenomenon that a minimal resistivity appears during annealing is attributed to hydrogen-related donors associated with the Si-H bond. In order to prove this, an experiment was carried out in which a H_2-FZ silicon ingot was pre-annealed at 700°C before irradiation. The Si-H bonds of the ingot were broken down during this annealing procedure. The annealing behavior of this ingot after irradiation is very similar to that of NTD Ar-FZ Si and did not exhibit a minimum resistivity during annealing.

The resistivity of either NTD Ar-FZ Si or NTD H_2-FZ Si reached the stable target value at an adequate annealing time and temperature, for example, 850°C for 2 hours.

Infrared Transmission Measurements

The transmitted intensity, I, of near-infrared light passing through a silicon sample is given by:

$$I = RI_o e^{-CX} , \tag{1}$$

where: I_o is the initial intensity of near-infrared light;
X is the thickness of the silicon sample;
C is the absorption coefficient of the silicon sample;
R is the net reflection coefficient from the front and back surfaces of the silicon sample.

The absorption coefficient, C, depends on the impurities or defects in the sample that can absorb or scatter near-infrared light at characteristic wavelengths.

The transmitted intensity and absorption coefficient are I_a and C_a before irradiation, respectively. After irradiation, these values are I_i and $C_a + C_i$, with i corresponding to different annealing temperatures, where C_i is the contribution of the defects (or impurities) produced during irradiation as well as annealing. Before annealing, i = 1 and the following relationship can be derived from eq 1:

$$\frac{C_i}{C_1} = \frac{LnI_a - LnI_i}{LnI_a - LnI_1} . \tag{2}$$

The results of substituting the experimental values, I_a and I_i into eq 2 are shown in figure 2. It can be seen from this figure that more than 90 percent of the defects due to irradiation disappeared during annealing at 450°C. However, as the annealing temperature continues to increase, the ratio C_i/C_1 increases rather than decreases. This is attributed to the hydrogen precipitate occurring after disruption of the Si-H bond.

Mid-Infrared Spectrum

The mid-infrared absorption spectrum of H_2-FZ Si has been measured at a number of annealing temperatures. Many peaks of this spectrum are related to Si-H bonds, such as the peaks at 2207, 2180, 2160, 2120, 1946 (cm^{-1})[1]. However, the peak at 2160 cm^{-1} is a new one which appeared when the annealing temperature was increasing. It reaches the maximum at 450°C and disappears at temperatures beyond 650°C. This behavior is shown in figure 3. This phenomenon appears to coincide with the resistivity changes and probably correlates to the postulated hydrogen-related donor.

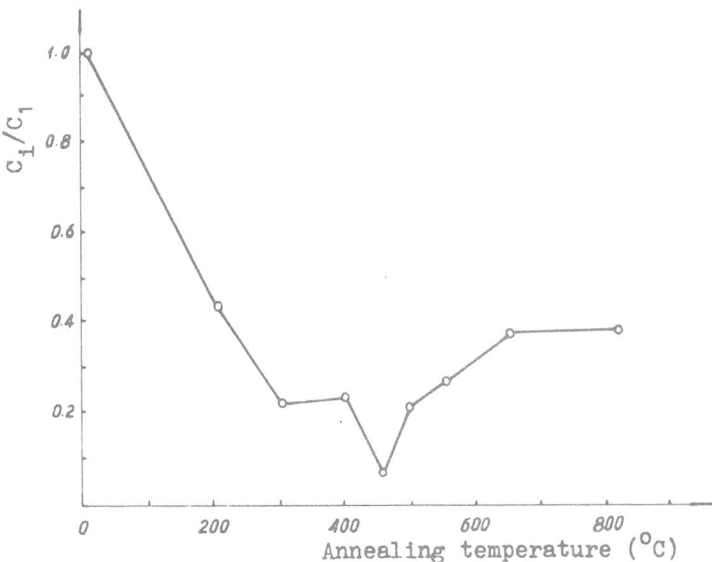

Fig. 2. Effect of annealing on the infrared transmission of H_2-FZ Si at a wavenumber range of 4000-10000 cm^{-1} and temperature of 300°K.

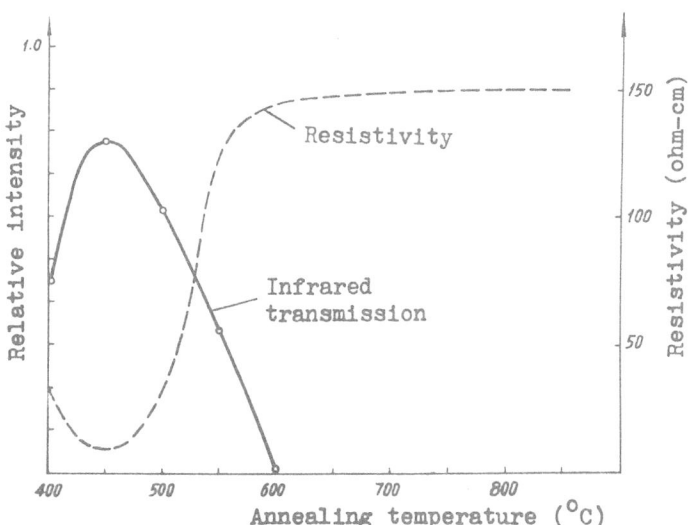

Fig. 3. Infrared transmission at 2160 cm^{-1} versus annealing temperature for a NTD H_2-FZ Si sample (0.5 h isochronal annealing).

All of the peaks attributed to Si-H bonds disappear at temperatures above 650°C. This indicates that all Si-H bonds have
been broken. The breakdown of Si-H bonds produces many hydrogen
precipitates in the silicon ingot. Since these precipitates will
cause hydrogen-induced defects[2] and adversely influence the
quality of silicon, they have to be eliminated. X-ray topography
of silicon samples and devices manufactured showed that this kind
of defect can be eliminated effectively by using slice annealing
techniques (850°C, 0.5 h).

Micro-Hardness

The hardness of NTD H_2-FZ Si was measured with a microhardom-
eter before and after irradiation. The root-mean-square error of
measurement was ±6.5 percent. The average value of hardness was
999 kg/mm^2 before irradiation and 1000.8 kg/mm^2 after irradiation.
The deviation between these values is within experimental error
and not considered significant. This indicates that the neutron
transmutation doping does not affect the hardness of H_2-FZ Si.

Other tests were also performed, such as the migration rate,
the lifetime of minority carriers, positron annihilation, etc.
The results obtained from these tests and analyses show that the
existence of hydrogen in NTD H_2-FZ Si during annealing does affect
these characteristics. However, most of these effects no longer
exist as the annealing temperature is increased above 650°C.
Although the effect of hydrogen precipitates caused by Si-H bond
breakdown is more significant, it can be eliminated by taking
suitable precautions. Except for this, most characteristics of
NTD H_2-FZ Si are identical with those of NTD Ar-FZ Si.

ELECTRICAL CHARACTERISTICS OF NTD H_2-FZ SILICON

Irradiation Fluence

The irradiation fluence received by a silicon ingot was determined with Al-Co alloy foils attached to the ingot. Figure 4
shows the distribution of the fluence for ten selected ingots.

The indicated length of the vertical lines in the figure
represents the deviations of the fluence along the axial direction
of the ingot. The difference for each ingot between the actual
and target fluence is within ±5 percent.

Accuracy for the Final Resistivity

After neutron doping, a certain number of samples were sliced
from each ingot to make rectifiers. After lapping-off the diffusion layer, the resistivities of the silicon substrate were mea-

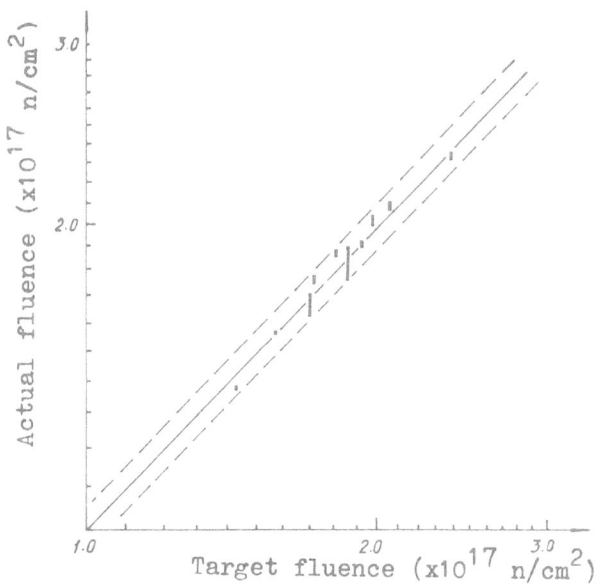

Fig. 4. Correspondence of the actual fluence and the target fluence.

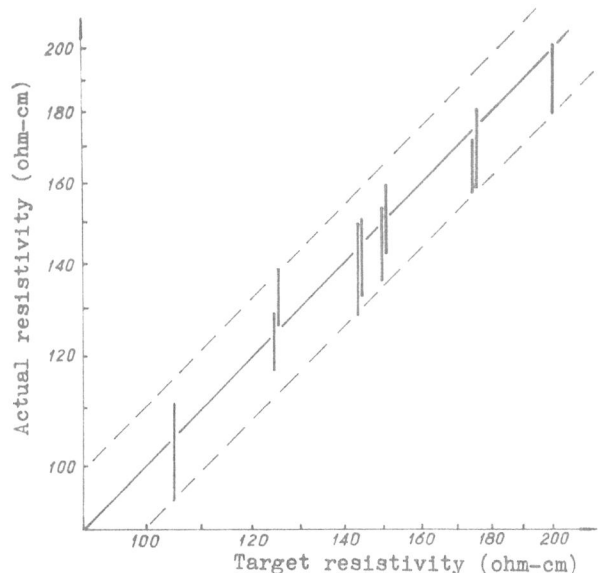

Fig. 5. Deviation between the measured actual resistivity and the target resistivity for NTD H_2-FZ Si.

Table 1. Thermal Stability of Resistivity of NTD H_2-FZ Si After
          Annealing

Number of samples	Target resistivity (ohm-cm)	Resistivity after annealing		$\dfrac{\rho_a - \rho_h}{\rho_h}$
		$T=850\,°C$ ρ_a (ohm-cm)	$T=1260\,°C$ ρ_h (ohm-cm)	
10	125	126.0	127.7	-1.3%
9	150	160.1	154.9	3.4%
10	200	207.6	204.9	1.3%

sured at five points using a four-point probe, taking the mean of
them as the slice resistivity. The final resistivity of that
ingot was obtained as the average of the slice resistivities. The
correlation between the final ingot resistivity and the target
ingot resistivity is shown in figure 5. The slice resistivities
deviate less than ±10 percent from the target resistivities.

Thermal Stability of the Resistivity

Since annealing of irradiation damage can occur during device
fabrication, a separate annealing is not required for NTD silicon
ingots. However, since device manufacturers are concerned about
the thermal stability, a study of resistivity stability after a
850°C anneal was made. Slices were taken from three unannealed
silicon ingots. These samples were annealed at 850°C for 2 hours
and their resistivities were measured. Then, these samples were
annealed again at 1260°C for 20 hours, and their resistivities re-
measured. The results are listed in table 1. The fact that the
resistivity differences between these two groups are very small
shows that the thermal stability of the samples annealed at 850°C
is very good.

Uniformity of Slice Resistivities

Eight to ten silicon slices were taken from each of the three
ingots measured above and annealed at 850°C; then the resistivities at five points were measured with a four-point probe. The
nonuniformity of the resistivities for each slice was calculated
as follows:

$$\Delta\rho = \frac{\rho_{max} - \rho_{min}}{\rho_{min}}. \tag{3}$$

The radial distribution of slice resistivities is shown in
figure 6. All of the nonuniformities are less than 7 percent.

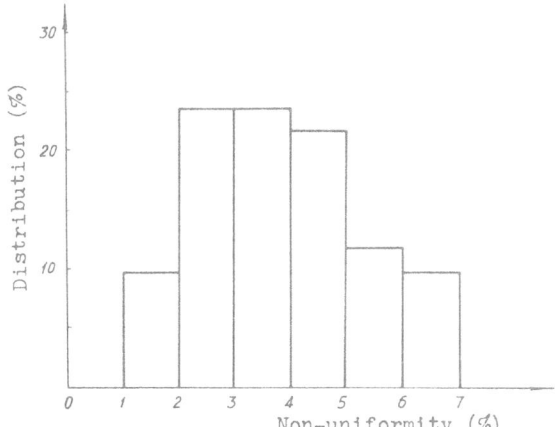

Fig. 6. Distribution of radial nonuniformities of slice resistivity.

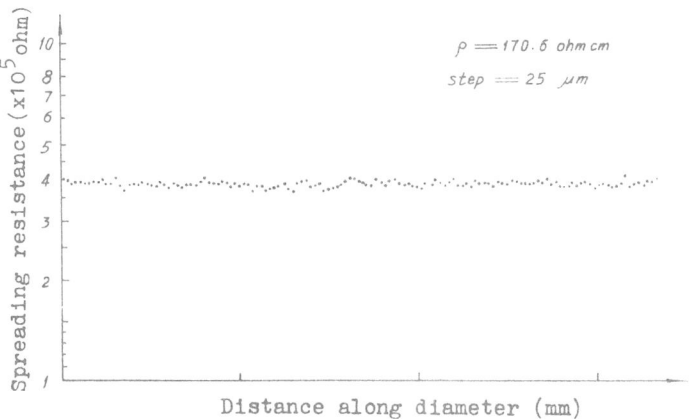

Fig. 7. Spreading resistance profile of NTD H_2-FZ Si slice.

Striations

Figure 7 shows a typical spreading resistance profile for a NTD H_2-FZ Si slice after annealing at 850°C for 2 hours. The nonuniformity of striation resistivity is defined as:

$$\varepsilon = \pm \frac{\rho_{max} - \rho_{min}}{\rho_{max} + \rho_{min}} \tag{4}$$

and was found to be ±5 percent.

From the results mentioned above, it can be seen that the existence of hydrogen in silicon does not affect the doping accuracy, the doping uniformity, and the thermal stability of the resistivity after doping.

FABRICATION OF HIGH POWER DEVICES

The fabrication of high power devices using the NTD H_2-FZ Si is carried out in the Institute of Rectifier in Xian. They fabricate rectifiers using 100-200 ohm-cm NTD silicon and thyristors using 120-150 ohm-cm NTD silicon. A separate annealing process to remove irradiation damage is not included in the fabrication schedule for these devices, and conventional technology is used in making these devices.

Rectifiers

Rectifiers were made from some slices cut from ten ingots of NTD H_2-FZ Si and one ingot of NTD Ar-FZ Si. The breakdown voltage characteristics of these devices are illustrated in figure 8. It can be seen that the measured values are higher than those obtained from conventional processing (solid line in the figure). This difference for NTD silicon has been reported previously[3]. Chou Zhongding has modified an empirical relationship for conventional processing (dashed line in the figure) to obtain better agreement with these NTD data[4].

Most of the measured minority carrier lifetimes were above 30 μs; the maximum value observed was 87 μs.

During the various diffusion treatments used in making devices, the slice resistivity was observed to remain stable.

Thyristors

High power thyristors were made from 125 slices using 120-150 ohm-cm NTD H_2-FZ Si and from 25 slices using 120-130 ohm-cm NTD Ar-FZ Si. The parameters of these thyristors are listed in table

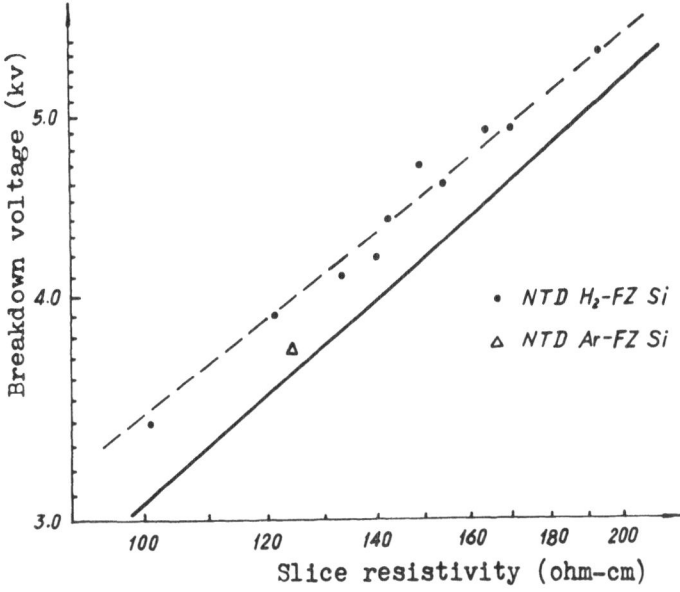

Fig. 8. Breakdown voltage versus slice resistivity. The solid line is the relationship expected for conventional processing.

Table 2. Electrical Characteristics of 50 mm Diameter Thyristors

Type of sample	Blocking voltage		Forward voltage drop		dV/dt (V/μs)	L (μm)
	forward (V)	reverse (V)	I=500A (V)	I=1500A (V)		
NTD H$_2$-FZ Si	3000	3000	0.70	2.0-2.1	900	193-220
NTD Ar-FZ Si	3000	3000	0.66	1.96	900	215

2. The table shows that the results obtained from NTD H$_2$-FZ Si are essentially the same as those obtained from NTD Ar-FZ Si.

CONCLUSIONS

    The following conclusions can be drawn from the experiments and analyses discussed above.

1. The behavior of H$_2$-FZ Si after neutron transmutation doping exhibits the effect of the existence of hydrogen in silicon

and differs significantly from Ar-FZ silicon after neutron transmutation doping during annealing. These differences include: passage through a minimum resistivity, type conversion, and differences in the infrared transmission spectrum. However, when the annealing temperature is above 650°C, the effect due to hydrogen in the silicon disappears.

2. The problems arising from the hydrogen-induced defects due to hydrogen precipitates at annealing temperatures above 600°C can be solved effectively by means of a slice annealing process.

3. For high-power device fabrication using the NTD H_2-FZ Si, a preprocessing annealing step is not necessary since the annealing occurs during the device fabrication. This simplifies annealing technology and reduces the cost. Meanwhile, it reduces the probability of contaminating the silicon and increases the device quality.

4. The main characteristics of NTD H_2-FZ Si are close to those of the more common NTD Ar-FZ Si. Both are excellent materials for high power devices.

REFERENCES

(1) Cui Shufan, Ge Peiwen, Zhao Yagin, Wu Lansheng, ACTA physica sinica, Vol. 28, No. 6 (1979) p. 791.

(2) Jiang Vailin, Sheng Shixiong, Xia Zhigang, ACTA physica sinica, Vol. 29, No. 10 (1980) p. 1283.

(3) A. Senes, Neutron Transmutation Doped Silicon, ed. by Jens Guldberg (1981) p. 339.

(4) Zhuo Zhongding, Power Electronics, No. 3 (1981) p. 54.

EXPERIENCE WITH NEUTRON TRANSMUTATION DOPED SILICON IN

THE PRODUCTION OF HIGH POWER THYRISTORS

Anders Alm, Gerhard Fiedler and Mirka Mikes

ASEA AB, Power Electronics Subdivision
Semiconductor Components Department
S721 83 Västerås, Sweden

ABSTRACT

High power thyristors have been processed, using NTD silicon ingots preannealed at 800°C and 1,200°C. Annealing at these temperatures produces different lifetime distributions as measured before processing into thyristors using a light-spot method and liquid rectifying contacts. The material annealed at 1,200°C shows lifetime variations which reflect the spiral patterns arising during the growth of the crystal. These lifetime variations were not eliminated by the processing but remained in fully-processed thyristors. The effect of these lifetime variations was to increase the leakage current in the thyristors at temperatures above room temperature. A deep level possibly responsible for this effect was detected by the DLTS method. Some methods are described which were found to be successful in diminishing these lifetime variations. Material annealed at 800°C did not exhibit these lifetime variations.

INTRODUCTION

Originally, Neutron Transmutation Doping was introduced as a means of improving the resistivity homogeneity over a silicon wafer[1]. Many of the disadvantages of conventional doping were avoided by this method and most of the electrical parameters characterizing a thyristor were improved[2].

However, NTD requires irradiation with neutrons, which damages the crystal, and a subsequent annealing by heating. It has been shown[3] that both the resistivity and the lifetime of the material are affected by the annealing procedure.

When manufacturing high power thyristors with NTD silicon, significant differences between materials annealed at different temperatures have been observed. Some annealing procedures have resulted in thyristors with higher leakage currents. To investigate the origin of this effect, we have performed experiments with silicon annealed at 1,200°C and at 800°C for 5 hours. The resistivity and lifetime were investigated in various stages of the thyristor processing sequence, in order to find out whether the effect was related to the starting material or induced by the subsequent high temperature processing steps.

EXPERIMENTS

Materials description

The starting silicon used in the present experiments had the following characteristics:
- n-type float-zoned NTD
- resistivity ~ 200 ohm-cm
The annealing was performed on the ingots after NTD irradiation, but before fabrication into thyristors.

The ingots were annealed at two different temperatures and designated as type A and type B to indicate annealing temperatures of 800°C and 1,200°C respectively.

Experimental methods

In the investigation the following types of measurements were made:
- Resistivity measurements
- Characterization of deep levels by DLTS
- Lifetime measurements

Resistivity measurements. The resistivity measurements utilized two techniques:
- four-point probe
- spreading resistance method

Characterization of deep levels by DLTS. This method was used only for qualitative purposes, to compare signal spectra from samples taken from the two different types of silicon. A conventional DLTS system[4] was used with a lock-in amplifier for signal averaging.

Lifetime measurements. The technique used is described in reference 5. Essentially, the photoresponse from the wafer was measured, when a laser beam was incident on the wafer surface.

The distribution of the photoresponse over the wafer was obtained by scanning the beam over the wafer surface using a system of vibrating mirrors. He-Ne lasers of two wavelengths were used: red light (0.6328 μm) and infrared light (1.15 μm). The experimental setup is pictured schematically in figure 1. The photoresponse from unprocessed wafers could be obtained by using a liquid rectifying contact (1% NH_4F solution) as described in reference 6. Due to the surface condition of our material, observations were limited to low voltages (< 50 V).

The following equations can be written for the photocurrent under different experimental conditions:

a)   for rectifying contacts:

$$I_{red} = f(\phi, L_p, s) \, \exp\left[\frac{W_{sc}}{L_p}\right] \qquad (1)$$

$$I_{IR} = q\alpha\phi \, (L_p + W_{sc}) \qquad (2)$$

b)   for a p-n structure:

$$I_{red} = f(\phi, L_p, s) \, \exp\left[\frac{W_{sc}}{L_p}\right] \qquad (3)$$

$$I_{IR} = q\alpha\phi \, (L_p + L_n + W_{sc}) \qquad (4)$$

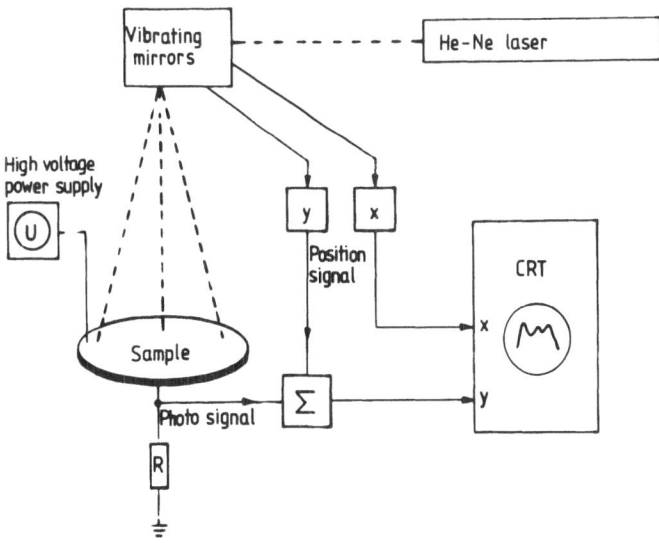

Fig. 1   Schematical picture of lifetime measuring equipment

c)   for a pnp-structure:

$$I_{IR} = q\alpha\phi \frac{L_p + L_n + W_{sc}}{1 - \left[\cosh(\frac{d - W_{sc}}{L_p})\right]^{-1}}$$ (5)

where

$I_{IR,red}$	is the photocurrent density response for infrared light and red light respectively.
$W_{sc} = \gamma(\rho V)^{\frac{1}{2}}$	is the width of the space charge region.
$\gamma = (2\kappa\varepsilon_0\mu_n)^{\frac{1}{2}}$	is a constant containing the dielectric constant and the mobility of electrons.
ρ	is the silicon resistivity
q	is the electronic charge
V	is the applied reverse bias voltage which is assumed to be very much larger than the thermal-equilibrium built-in potential.
$L_{p,n} = (D_{p,n}\ \tau_{p,n})^{\frac{1}{2}}$	is the diffusion length of holes and electrons respectively.
$D_{p,n}$	is the diffusivity of holes and electrons respectively.
$\tau_{p,n}$	is the lifetime of holes and electrons respectively.
ϕ	is the flux of incident photons per unit area.
α	is the absorption coefficient for the infrared light
s	is the surface recombination velocity
$f(\phi, L_p, s)$	is a function independent of the applied reverse voltage V [9].

The effect of surface recombination was neglected for the infrared measurements.

From equations (1) and (3) it can be concluded that the signal I_{red} is difficult to interpret since the ratio of resistivity and lifetime enters the expression. Thus variations in one of the entities can be detected only if the behaviour of the other is known.

The expressions (2) and (4) are much simpler, being sums of L_p, L_n and W_{sc}. For the IR-laser light it is possible to separate lifetime and resistivity variations for the following reasons:

In the resistivity range used ($\sim$ 200 ohm-cm) the voltage 10 V yields $W_{sc} \approx 35$ µm (part of which is due to the diffusion potential). For the lifetime level in the starting material (≈ 100 µs) one obtains $L_p \approx 350$ µm. For low voltages it is therefore reasonable to neglect W_{sc} in comparison with L_p (or L_n) i.e. signal variations are interpreted as being due to lifetime variations.

For higher voltages (as can be used for p-n junctions) the term W_{sc} starts to compete with the terms L_p and L_n. For example, V = 2 kV yields $W_{sc} \approx 330$ µm. By performing the measurements at low and high voltages it is therefore possible in this case to distinguish between resistivity and lifetime variations. For pnp-systems the I_{IR} signal follows an expression containing

$$\left[\cosh(\frac{d-W_{sc}}{L_p}) \right] - 1$$

due to hole injection from the p-base. Below 2 kV this expression varies slowly and therefore the signal variation in this case reflects the lifetime variations at low voltage and a composite of resistivity and lifetime at higher voltages.

To verify experimentally the arguments above some measurements were performed on an unprocessed wafer. Figure 2 is a four-point probe measurement of the resistivity variation over the wafer. Figure 3 is the photocurrent signal variation of the wafer at a low voltage ($\sim$ 15 V). It is clearly seen that the photocurrent signal is uniform even though considerable resistivity variations are present. This supports the assumption that the photocurrent signal in this case is independent of resistivity and depends on lifetime.

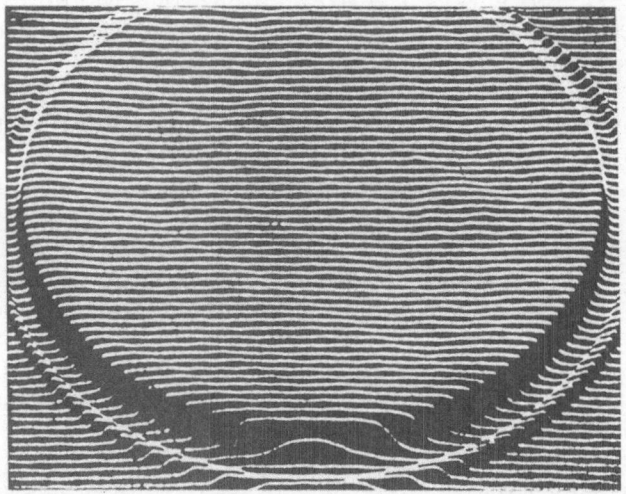

Fig. 2   Four-point probe measurements of relative radial
          resistivity variations of three n-type testwafers.

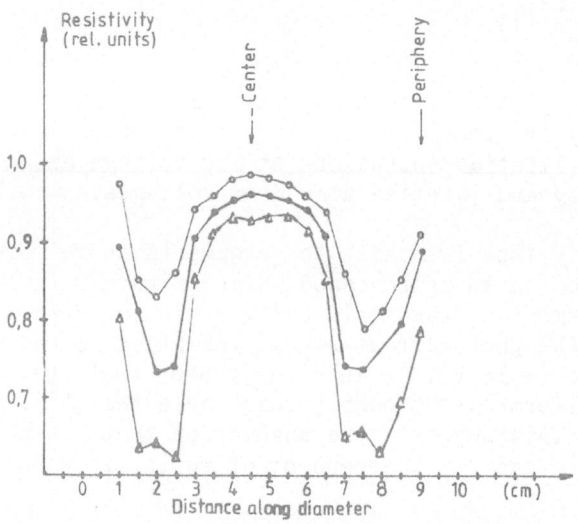

Fig. 3   Photoresponse map of one of the test wafers in
          figure 2.

INVESTIGATION OF UNPROCESSED MATERIAL

Measurements were performed on unprocessed material of
type A (800°C preanneal) and type B ($1,200^\circ$C preanneal).

The resistivity was checked using a four-point probe measurement. It was found to be within specifications from the supplier.
A slight tendency for variations over the wafer could be determined
statistically over a number of wafers as is seen in figures 4 and 5,
but generally the variations over single wafers were negligible.

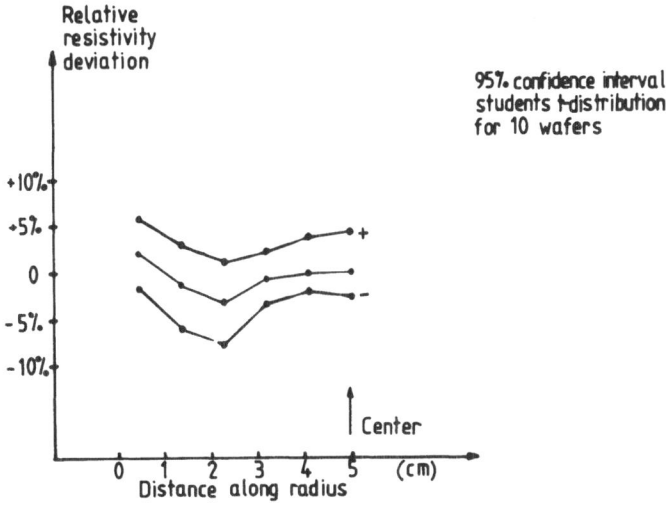

Fig. 4   Statistically obtained resistivity variation over type
          A wafers, measured with a four-point probe.

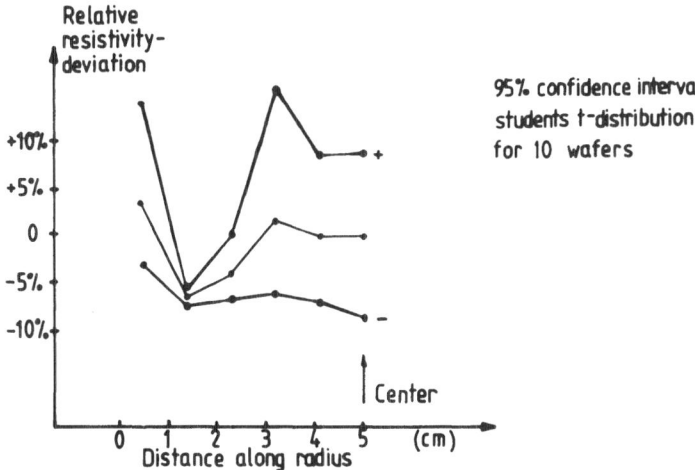

Fig. 5   Same as figure 4, but for type B wafers.

Figures 6 and 7 show the photoresponse mappings for two typical wafers of each type. The type A material has constant signal over the wafer whereas the type B material shows a considerable variation of the wafer. Since the measurements were performed at ≈ 10 V, these signal variations are interpreted as lifetime variations. It can be seen that the lifetime distribution

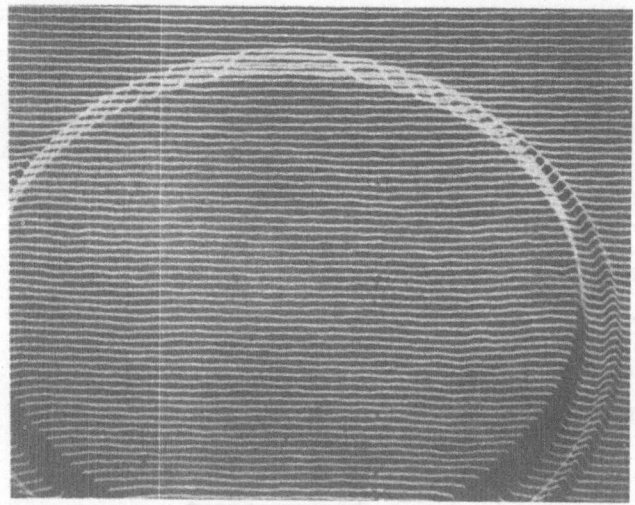

Fig. 6   Photoresponse map of a typical type A wafer, before processing using IR light.

Fig. 7   Photoresponse map of a typical type B wafer, before processing using IR light.

pattern has a spiral like form, similar to that for swirls and
striations, which is tentatively ascribed to the rotation of the
ingot during crystal growth.

INVESTIGATION OF MATERIALS DURING PROCESSING

During processing the silicon wafers are subjected to a
number of high temperature diffusion and oxidation steps. In
particular they are subjected to one phosphorous and one boron
diffusion towards the end of the processing sequence. Both of
these diffusions are known to have a gettering effect on heavy
metal impurities.

Figure 8 shows the photoresponse signal of a typical type B
wafer after phosphorous diffusion, using infrared light. For
the voltage used (750 V) large areas of the wafer show no
photoresponse at all. Figure 9 is the signal map of the same
wafer at 2 kV. Figure 10 shows the photoresponse over one linear
cut across the wafer, for several voltages. Figure 11 shows a
spreading resistance measurement over the same wafer.

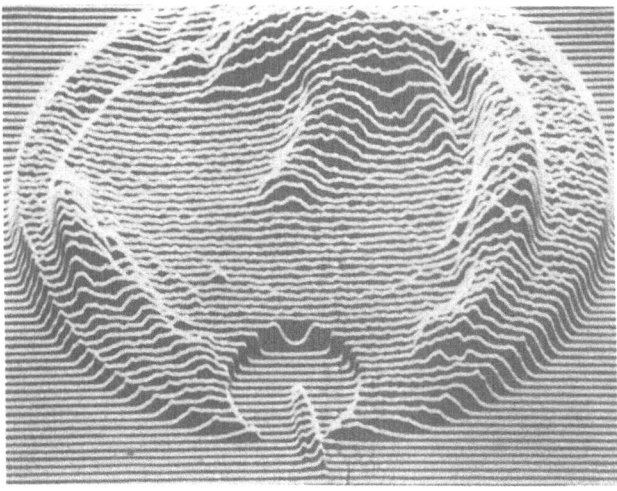

Fig. 8   Photoresponse map of a typical type B wafer after
         phosphorous diffusion, under a reverse bias of 750 V
         using IR light.

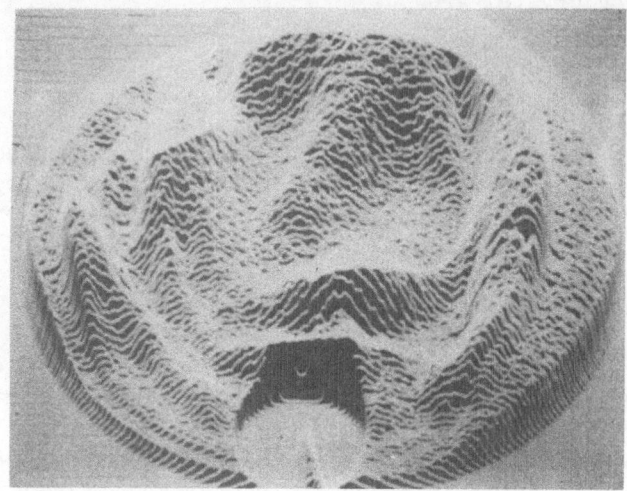

Fig. 9   Photoresponse map of same wafer as in figure 8 but
          with reverse bias 2 kV using IR light.

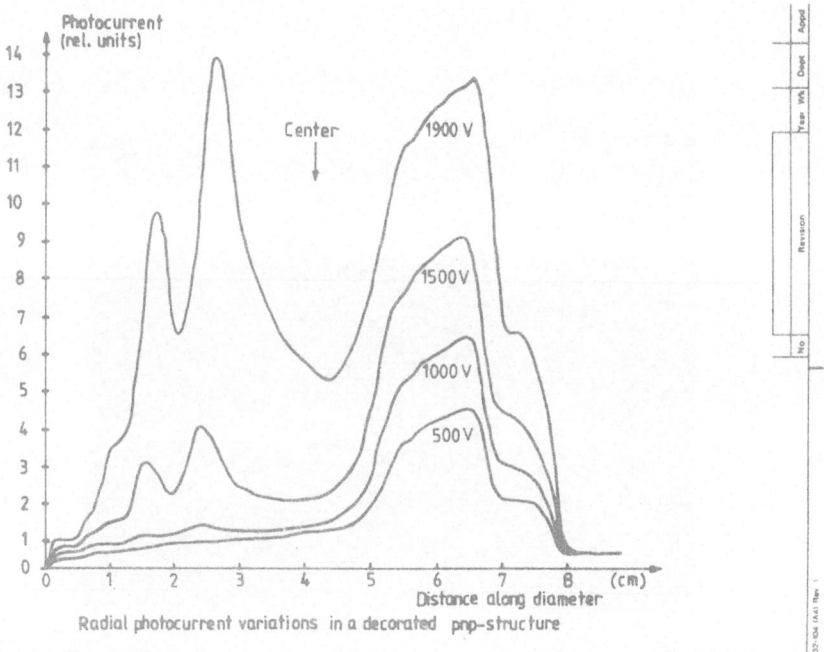

Fig. 10   Photoresponse at different voltages along a cut across the
          same wafer as in figures 8 och 9 using IR light.

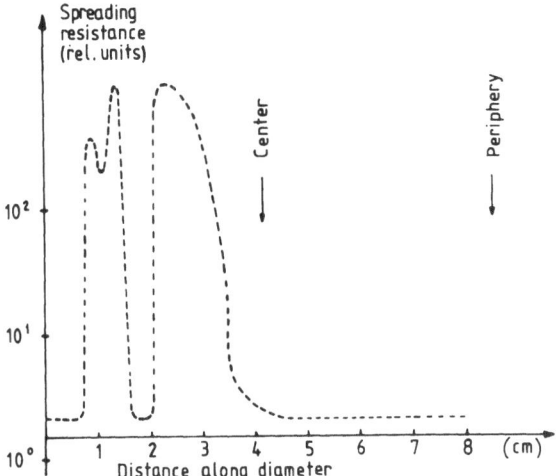

Fig. 11 Spreading resistance measurement along the same cut as in figure 10.

From these figures and eq. (5) it appears that the signal obtained at lower voltages reflects the lifetime and its variations. Thus we have a low lifetime area in the wafer, and an area with a higher but varying lifetime. The signal obtained at higher voltages in the low lifetime area is also varying. Its connection with resistivity variations is demonstrated by figure 11 and also by its voltage dependence. Its rapid rise at higher voltages reflects the fact that the space charge region approaches the opposite p-layer, which results in the expression:

$$\cosh\left[\left(\frac{d-W_{sc}}{L_p}\right)\right]^{-1}$$

approaching unity in eq. (5). It should be noticed that the resistivity and lifetime variations follow a spiral like pattern similar to that found in the starting material. The same photoresponse patterns are found in type B material after boron diffusion and in a finished thyristor (i.e. after gold diffusion), see figure 12 and 13.

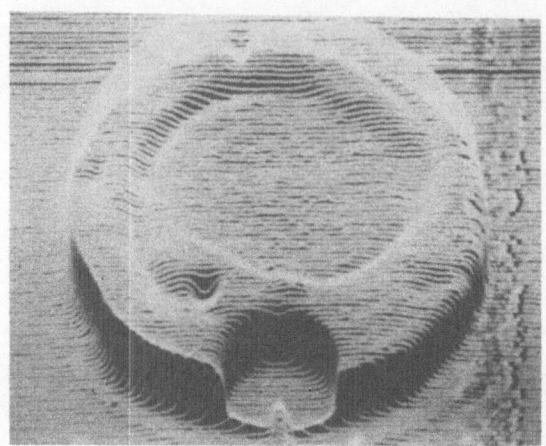

Fig. 12   Photoresponse map of a typical type B wafer after
boron diffusion using IR light.

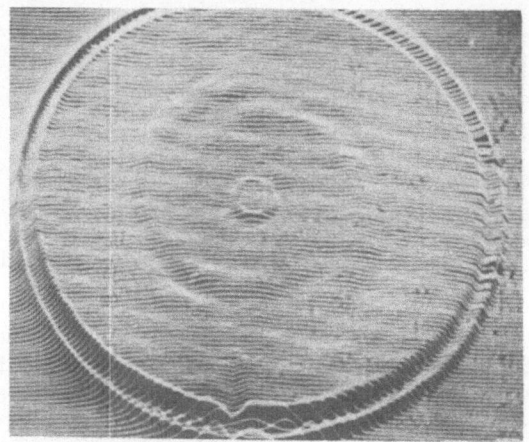

Fig. 13   Photoresponse map of a typical type B wafer
(finished thyristor) using IR light.

In type A material on the other hand, the photoresponse
signal is constant over the wafer after all these process
steps, see for example figure 14 and 15.

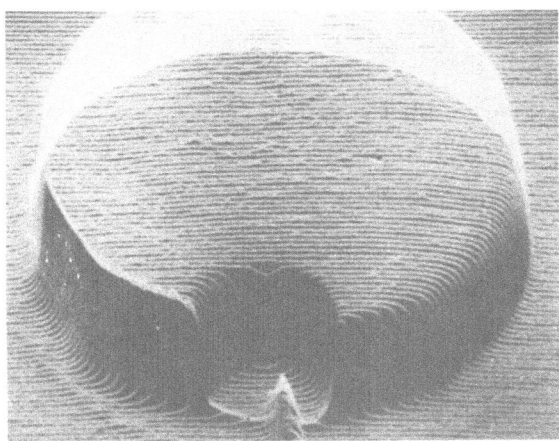

Fig. 14   Photoresponse map of a typical type A wafer after
boron diffusion using IR light.

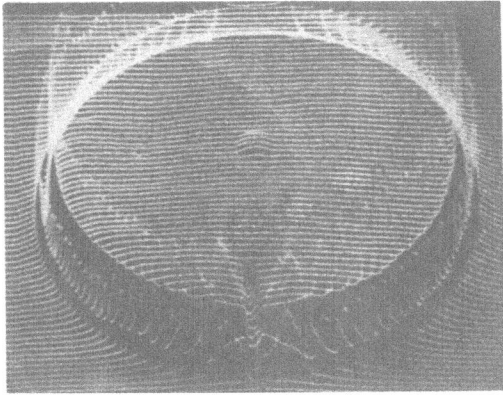

Fig. 15   Photoresponse map of a typical type A wafer,
finished component using IR light.

ELECTRICAL EFFECTS

The main difference of behaviour between thyristors made
from type A and type B materials is that the latter have intolerably high leakage currents. The effect is most pronounced
above room temperature but is evident at room temperature.
Figure 16 and 17 demonstrate the effect.

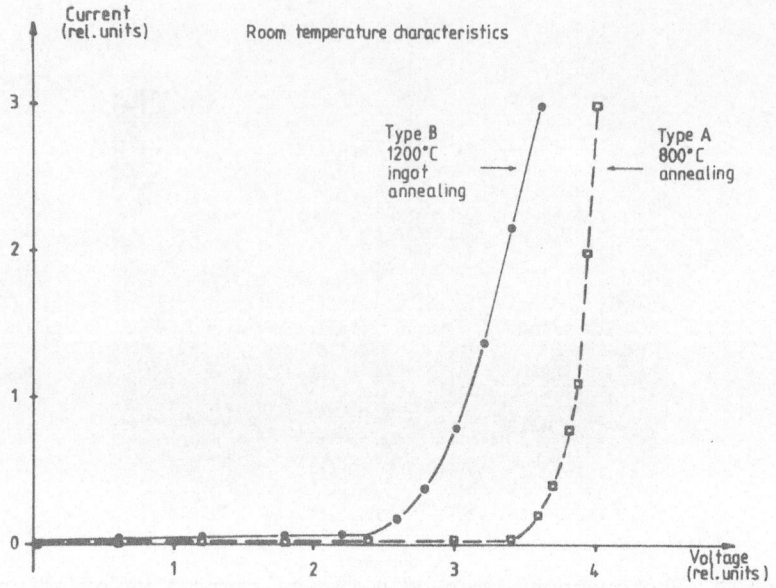

Fig. 16   Room temperature current voltage characteristics of
type A and type B thyristors.

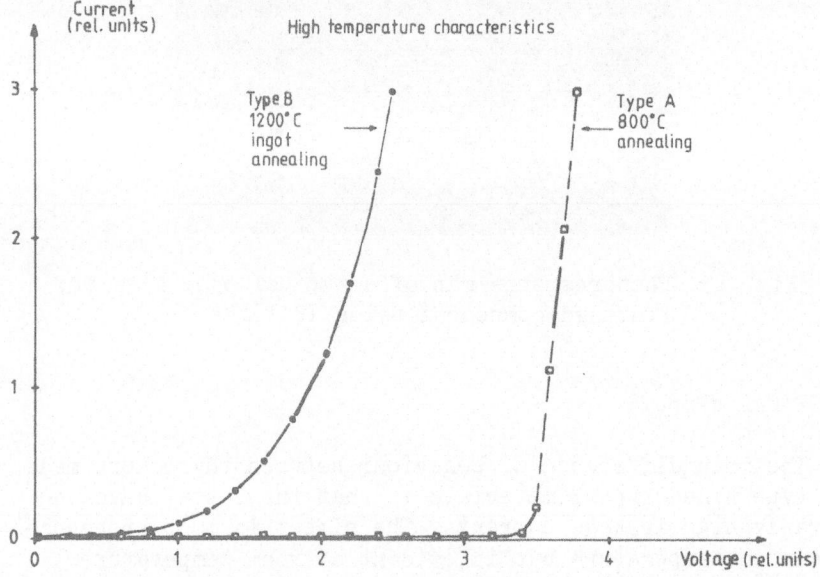

Fig. 17   High temperature current voltage characteristics of type A
and type B thyristors at 100°C.

Figure 18 shows the DLTS spectra for type A and type B materials in a finished thyristor. The essential difference between the two materials is the existence of a peak above room temperature, present only in the type B material. The thermal characteristics of this level implies that it is very deep, i.e. centered near the middle of the band gap. This level can be connected with the leakage current in the following way.

The expression for the generation leakage current is

$$I = qW_{sc}N_T (e_n^{-1} + e_p^{-1})^{-1} \qquad (6)$$

where:

$e_{n,p}$ are the thermal emission rates for electrons and holes respectively and are given by:

$$e_{n,p} = c_{n,p} \, N_{c,v} \exp \left(- \Delta E_{n,p}/kT \right) \qquad (7)$$

where:

$c_{n,p}$ is the electron or hole capture coefficient

$N_{c,v}$ is the density of states in the conduction or valence band

$\Delta E_{n,p}$ are the energy distances from the trap to the conduction and valence bands respectively.

N_T is the concentration of deep-level traps.

The temperature dependence of the leakage current is contained essentially in the term $(e_n^{-1} + e_p^{-1})^{-1}$. For the generation current to be high and have a strong temperature dependence, it is necessary that both e_n and e_p be large. Presuming that $c_nN_c \approx c_pN_v$, this implies that $\Delta E_n \approx \Delta E_p$, i.e. the level is centered near the middle of the band gap. This agrees with the characteristics of the deep level observed in the DLTS spectrum in figure 19. Therefore is seems probable that the leakage current behaviour of the components is controlled by this deep level.

DISCUSSION

Experimentally, the following properties are found in the type B NTD material annealed at 1,200°C:
- lifetime variations in unprocessed material
- lifetime and resistivity variations in processed material

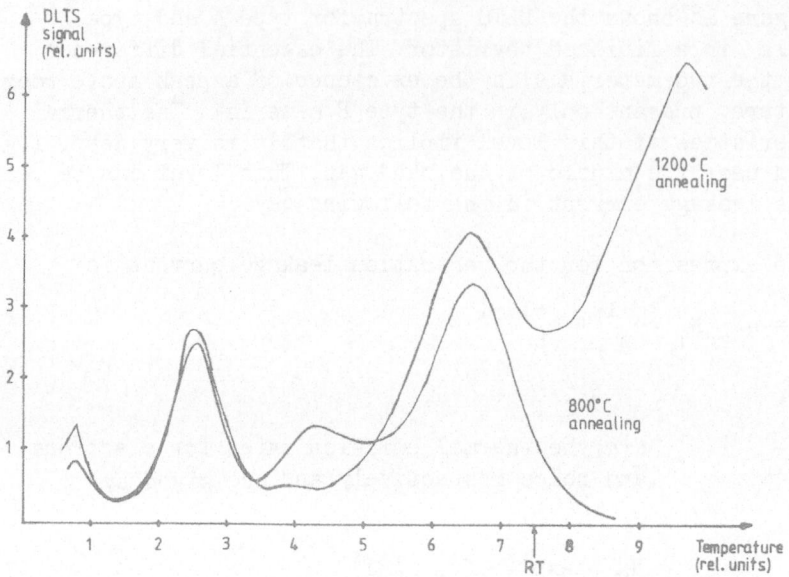

Fig. 18   DLTS spectra from finished type A and type B thyristors.
          Room temperature is indicated by the arrow marked RT.

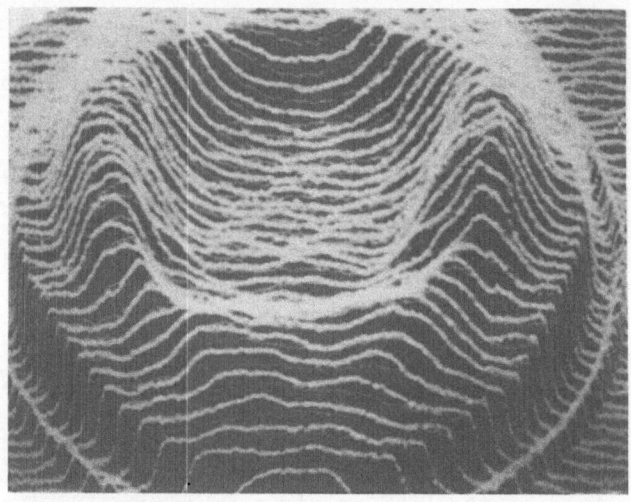

Fig. 19   Photoresponse map of a type B wafer (starting material)
          before HCl oxidation using IR light.

- the presence of a deep level not present in type A material.

    It seems probable that the deep level and the lifetime
variations have a common physical cause and that the deep level
is responsible for the increased leakage currents in type B
thyristors, as compared to type A thyristors. There are three
possible origins for this defect:

1. A defect, introduced in the crystal growing process, is converted to an electrically-active form during the high temperature anneal.

2. The defect is the result of a reaction between a chemical impurity (oxygen, carbon) and irradiation produced defects that takes place at $1,200^{\circ}$C but not at 800°C.

3. The defect is the result of complexing of irradiation produced defects.

Case 1 seems improbable because there are no obvious differences between type A and type B ingots of starting material. Although we cannot distinguish between cases 2 and 3 the former seems to be the more probable, due to the fact that the defect distribution is similar in form to the striated distribution of carbon and oxygen in silicon. However, the limited information on the as received material characteristics precludes any unambiguous conclusion concerning the origin of the defect.

The resistivity and lifetime variations in material exposed to gettering processes during fabrication indicated the following:
- the defects present in the type B starting material could be decorated in the subsequent processing.
- once decorated, the defects were stable enough to withstand gettering by phosphorous and boron.

Essentially, there are two techniques for eliminating the influence of the defects on the leakage current of the final thyristors. One method is to getter the defects before processing. The other method is to prevent decoration of the defect during processing. A number of experiments were made to evaluate these techniques.

HCl-oxidation

The effect of this process is to getter oxygen out of silicon and to eliminate stacking faults[7]. Figures 19 and 20 show the photoresponse mappings of a type B wafer before and after an HCl oxidation at $1,240^{\circ}$C. It is evident that the oxidation has removed the defects responsible for the lifetime variations.

Phosphorous gettering

The effect of this treatment is to remove both fast diffusers and crystal defects from silicon. The results of our experiments (not shown here) are identical with those of the HCl oxidation. Also, in both cases the leakage currents were reduced drastically after the treatment.

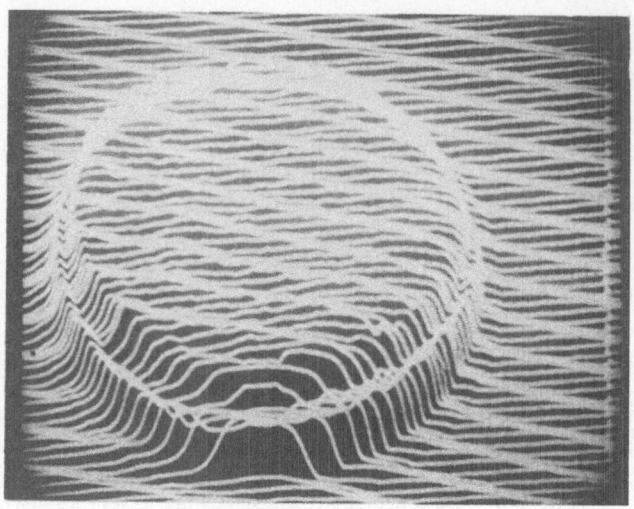

Fig. 20 Same thyristors as in figure 19, but after HCl-
         oxidation using infrared light.

Preprocessing cleaning treatment

Type B wafers were treated with a very weak HNO_3-HF etch,
and subsequently processed to thyristors. The purpose of this
etch was to remove contamination on the wafer surface that
conventional procedures would not remove. Control samples with
standard surface cleaning were processed simultaneously.
Figure 21 shows the I-V characteristics of diodes made from
wafers cleaned these two ways. It is evident that the etching
treatment has successfully removed the contamination originally
present at the silicon surface.

CONCLUSIONS

The annealing of ingots of NTD-doped silicon at $1,200^{\circ}C$
results in the presence of a defect in the materials. This defect
can be decorated during subsequent processing to the point where
substantial lifetime and resistivity variations occur, causing
intolerably high leakage currents in thyristors. By appropriate
gettering or cleaning methods, however, these adverse effects
can be avoided.

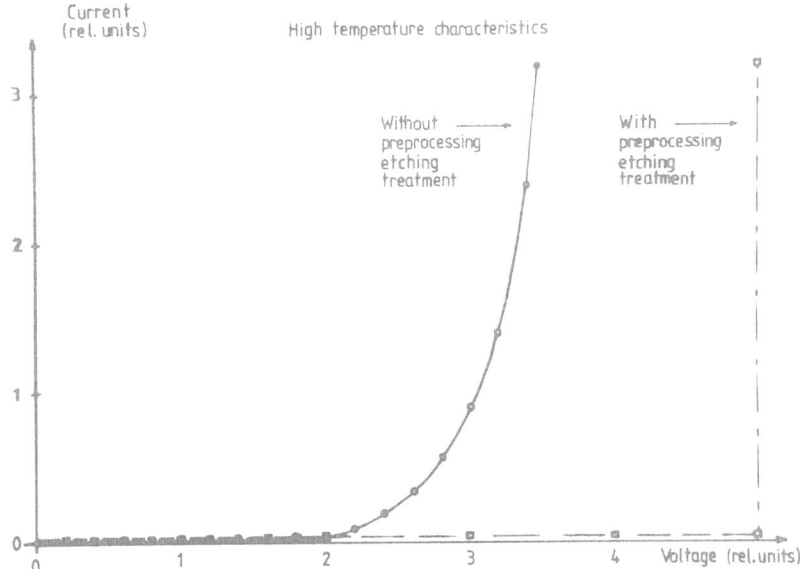

Fig. 21   Current voltage characteristics of type B components
at 100°C, with and without etching the starting material.

REFERENCES

1. H. Herrmann, H. Hertzer and E. Sirtl, "Festkörperprobleme XV"
   (Advances in Solid State Physics) pp. 279-316, Vieweg o Sohn,
   Wiesbaden (1975)
2. A. Senes, Proc. of the 3rd International Conference on NTD
   doping of Silicon in Copenhagen, Aug. 27-29, 1980, Plenum
   Press, N.Y. (1981)
3. H. Hertzer, Proc. of the 3rd International Conference on
   NTD doping of silicon in Copenhagen, Aug. 27-29, 1980,
   Plenum Press, N.Y. (1981)
4. D.V. Lang, J. Appl. Phys. 45, (1974) 3014; 45, (1974) 3023
5. O. Engström, B. Drugge and P.-A. Tove "Laser scanning
   techniques for the investigation of power devices" in
   "Lifetime Factors in Silicon", American Society for testing
   and materials Publ. ASTM STP 712, (1980), pp. 239-250

6. B. Drugge and E. Nordlander, IEEE transactions on electron
   Devices 27 (1980) pp. 2124-27.
7. K. V. Ravi, "Imperfections and impurities in semiconductor
   silicon", Wiley o Sons New York 1981, p. 334
8. Ibid p. 343
9. O. Engström, B. Drugge and P. A. Tove, Physica Scripta, Vol. 18,
   1978, p. 357.

TRANSIENT CURRENT SPECTROSCOPY OF NEUTRON IRRADIATED SILICON

J. W. Farmer and J. C. Nugent

University of Missouri
Research Reactor
Columbia, MO   65211

and

Department of Electrical Engineering
University of Missouri-Columbia
Columbia, MO   65211

ABSTRACT

Deep levels in semiconductors have been extensively studied
using capacitance transients. This technique is referred to as Deep
Level Transient Spectroscopy (DLTS). A similar technique using
transient currents is called Transient Current Spectroscopy (TCS).
The basis of both DLTS and TCS are discussed, and some advantages
of TCS over DLTS are pointed out. Some examples of past applications of DLTS to neutron irradiated silicon are given. The present
application of TCS to defects in silicon is presented. Defect
energy levels, capture cross sections and spatial profiles have
been determined using TCS.

INTRODUCTION

Transient Current Spectroscopy (TCS) is a simple variation of
the standard Deep Level Transient Spectroscopy (DLTS) technique.
DLTS has become a very popular technique for studying deep levels
(i.e., defects) in semiconductors. The popularity arises in part
from the fact that DLTS is relatively simple to use. It is possible
to obtain unique "signatures" of deep levels and DLTS is a very
sensitive technique. The DLTS technique utilizes the properties
of semiconductor junctions for the study of the defects. The two
junction properties most frequently used are capacitance and cur-

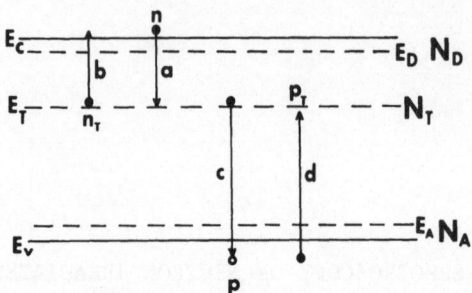

Fig. 1. Schematic of carrier capture and thermal emission
         processes at deep levels in a semiconductor (see glossary
         of terms in text).

rent. Traditionally, DLTS refers to studies in which capacitance
transients are used. When current transients are used the technique is generally referred to as TCS. While the two methods (capacitance and current) are fundamentally equivalent, there are experimental differences which often make the distinction convenient.
Thus, traditional convention will be used here in that DLTS will
refer to capacitance transient studies and TCS will refer to current transient studies.

     This paper will include a discussion of the principles upon
which the DLTS and TCS techniques are based. A few examples of the
past applications of DLTS to neutron irradiated Si are presented.
The results of our application of TCS to the study of defects in
Si are discussed. Finally, potential applications of DLTS (TCS)
to NTD-Si are presented.

BASIS OF DLTS AND TCS

Defect Properties

     From a technological point of view, the effect of defects on
the electronic structure of a semiconductor is by far the most
significant defect property. The effect of defect energy levels
on the electronic properties of a material can be described in
terms of thermal emission and capture rates of free carriers at
the defect levels (see Fig. 1). A glossary of the notation used
in this work follows:

n = concentration of electrons in conduction band (cm^{-3})
p = concentration of holes in valence band (cm^{-3})
N_D = concentration of donors (cm^{-3})
N_A = concentration of acceptors (cm^{-3})
N_T = concentration of defect energy levels (cm^{-3})
n_T = concentration of defect energy levels occupied by
     electrons (cm^{-3})
p_T = concentration of defect energy levels occupied by
     holes (cm^{-3})
E_C = energy of conduction band edge (eV)
E_V = energy of valence band edge (eV)
E_T = energy of defect energy level (eV)
E_F = energy of Fermi level (eV)
c_n = capture coefficient of defect for electrons (cm^3/sec)
c_p = capture coefficient of defect for holes (cm^3/sec)
e_n = emission coefficient for electrons at N_T (sec^{-1})
e_p = emission coefficient for holes at N_T (sec^{-1})
σ_n = capture cross section of defect for electrons (cm^{-2})
σ_p = capture cross section of defect for holes (cm^{-2})
g = degeneracy of trap energy level

The analysis that follows is based on Ref. 1. Other useful
papers are Refs. 2 through 5. The rate equations for the four
processes a, b, c and d in Fig. 1 are

$$-dn/dt = a - b = c_n n p_T - e_n n_T \tag{1}$$

$$-dp/dt = c - d = c_p p n_T - e_p p_T \tag{2}$$

$$dn_T/dt = (a - b) - (c - d)$$

$$= (c_n n + e_p)p_T - (c_p p + e_n)n_T \quad . \tag{3}$$

Note that

$$N_T = p_T + n_T \quad . \tag{4}$$

In this general form, the rate equations form a system of coupled
non-linear differential equations. However, if one considers the
region within the depletion width of a junction where $n = p = 0$,
the following important equation is obtained

$$dn_T/dt = -(e_n + e_p)n_T + e_p N_T \quad . \tag{5}$$

Eq. (5) is a simple first order linear differential equation which,
given the initial (or boundary) conditions, is readily solved.
Eq. (5) together with Eq. (4) provide the basic framework of DLTS.

Consider the following initial condition. In an n-type depletion region, for $t = 0$ let $n_T(0) = N_T$, i.e., all the traps are

filled with electrons. This initial condition corresponds to a majority carrier trap in n-type material. The solution of Eq. (5) gives

$$n_T(t) = N_T \left[\frac{e_p}{e_n + e_p} + \frac{e_n}{e_n + e_p} \; e^{-(e_n + e_p)t} \right] . \tag{6}$$

For traps above the middle of the band gap, it is generally true that $e_n \gg e_p$, thus Eq. (6) becomes

$$n_T(t) = N_T \; e^{-e_n t} . \tag{7}$$

Eq. (4) can be used to find the hole occupation, thus

$$p_T(t) = \frac{N_T \; e_n}{e_n + e_p} \left[1 - e^{-(e_n + e_p)t} \right] . \tag{8}$$

Similar solutions exist for minority carrier traps and for p-type material.

To determine the temperature dependence of the thermal emission rate, the probability, P, that a trap is occupied with an electron can be defined in terms of Fermi statistics as

$$P = n_T/N_T = \left[1 + g^{-1} \; e^{(E_T - E_F)/kT} \right]^{-1} . \tag{9}$$

In steady-state, the total emission rate must equal the total capture rate, therefore

emission rate   capture rate

$$e_n P \qquad = \qquad c_n n(1 - P)$$

or

$$e_n = c_n n \; g^{-1} \; e^{(E_T - E_F)/kT} .$$

Defining the capture coefficient in terms of the capture cross section (σ_n), $c_n n = n \langle v_n \rangle \sigma_n = N_C \; e^{-(E_C - E_F)/kT} \langle v_n \rangle \sigma_n$, yields (assuming $g = 1$)

$$e_n = N_C \sigma_n \langle V_n \rangle \; e^{-(E_C - E_T)/kT} \tag{10}$$

and by similar arguments

$$e_p = N_V \sigma_p \langle V_p \rangle \; e^{-(E_T - E_V)/kT} , \tag{11}$$

where N_c = effective density of states at conduction band edge
      N_v = effective density of state at valence band edge
      $\langle v \rangle$ = average carrier thermal velocity = $(3kT/m^*)^{1/2}$
      k = Boltzman constant
      T = temperature
      m^* = effective mass of the carrier.

For majority traps in n-type material, Eq. (7) states that $n_T(t)$ is a decaying exponential and, from Eqs. (10) and (11), the rate of decay depends exponentially on the energy difference between the trap and the corresponding band edge and on temperature.

Experimental Technique

The basic DLTS technique involves establishing the initial conditions, monitoring the decay process and varying the temperature. The initial conditions can be set by electronically pulsing the diode. When a reversed biased diode is pulsed to zero bias, the majority carriers will collapse the depletion width and fill majority carrier traps (see Fig. 2a), thus establishing the desired boundary condition of $n_T(0) = N_T$ for electron majority traps.

Given an established boundary condition, the decaying trap, $n_T(t)$, must be monitored. There are two junction properties which readily allow this to be done: capacitance and current. The capacitance as a function of time is given by $C(t) = A\varepsilon_s/W(t)$ where A is junction area, ε_s is semiconductor permittivity and $W(t)$ is the depletion width as a function of time. With no traps, $W = [2\varepsilon_s(V_{bi} + V)/qN_B]^{1/2}$ where V_{bi} is the built-in junction voltage, V is the applied voltage, q is the elementary charge and $N_B = N_D N_A/(N_D + N_A)$. For p^+n junctions, electron (majority carrier) traps will result in $N_B \rightarrow N_B - n(t)$. For n^+p junctions, electron (minority carrier) traps will result in $N_B \rightarrow N_B + n(t)$. Thus, for electron traps

$$C(t) = A[q\varepsilon_s N_B/2(V_{bi} + V)]^{1/2} \cdot \left[1 \mp \frac{n_T(t)}{N_B} \right]^{1/2} ,$$

the sign is determined by whether the electrons are majority carriers (−) or minority carriers (+). For $n_T(t) \ll N_B$, then

$$C(t) = C_\infty \left[1 \mp \frac{n_T(t)}{2N_B} \right] \tag{12}$$

where $C_\infty = A[q\varepsilon_s N_B/2(V_{bi} + V)]^{1/2}$. An identical result holds for hole traps where $n_T(t)$ is replaced by $p_T(t)$. Again, the sign is

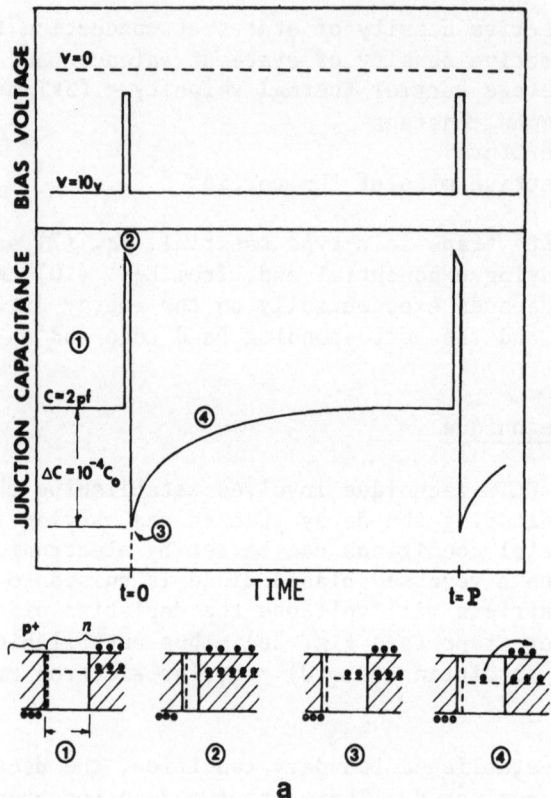

a

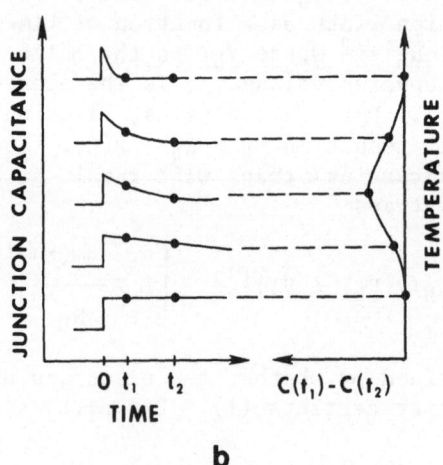

b

Fig. 2.   a) Diode bias and junction capacitance as a function of
              time.
          b) Capacitance difference signal as a function of
              temperature.

determined by whether the holes are majority carriers (-) or minority carriers (+).

For junction current,

$$i(t) = i_n(t) + i_p(t) + i_d(t) \qquad\qquad (13)$$

where $i_n(t) = A \int_0^W q(dn/dt)dx$ electron current

$\qquad i_p(t) = A \int_0^W q(dp/dt)dx$ hole current

$\qquad i_d(t) = A \int_0^W q(dn_T/dt)(x/W)dx$ Maxwell displacement current

For electron majority traps and for uniform doping and trap distributions,

$$i(t) = \frac{AqW}{2} (e_n n_T + e_p p_T) = \frac{AqW}{2} (e_p N_T + (e_n - e_p)n_T(t))$$

or for $e_n \gg e_p$ with $t \ll 1/e_n \cdot \ln\left(\dfrac{e_n}{e_p}\right) \approx \dfrac{E_g}{e_n kT}$, where E_g = band gap

$$i(t) = \frac{AqW e_n n_T(t)}{2} \; . \qquad\qquad (14)$$

In the above derivation, the time dependence of W has been ignored. If the condition $N_T \ll N_B$ is not met, then $W(t)$ must be explicitly included in the above analysis.

The methods for establishing the boundary conditions for the solution of the differential equation for $n_T(t)$ have been described. The method by which a DLTS spectrum is obtained can now be seen qualitatively in Fig. 2b. In this example, the difference in capacitance at two times is monitored as a function of temperature. At low temperatures, none of the traps thermally empty and the difference signal is zero. As the temperature is increased the traps begin to empty and a difference signal is observed. At even higher temperatures, all of the traps will empty before t_1 and again the difference signal will be zero. Thus the changing of temperature will give rise to a "peak" which is characteristic of the defect involved. By varying t_1 and t_2 for subsequent temperature scans, the defect peak will shift to different temperatures.

The relation between the peak signal observed in a temperature scan and the defect properties is seen in the following manner. For the case of capacitance measurements, the signal observed, S_c, is

$$S_c = C(t_1) - C(t_2) = \frac{C_\infty N_T}{2N_B} (\exp(-e_n t_2) - \exp(-e_n t_1)) \quad .$$

At the peak of the signal, the following relationships are true,

$$\frac{dS_c}{dT} = \frac{de_n}{dT} \cdot \frac{dS_c}{de_n} = 0 \quad . \tag{15}$$

Factoring out the explicit temperature dependence of the emission rate given in Eq. (10) we have for Si

$$e_n = 3.518 \times 10^{21} \, \sigma_n \, T^2 \exp(-(E_C - E_T)/kT) \tag{16}$$

where the effective mass of the electron is assumed to be $1.08 \, m_e$. It is clear from Eq. (16) that $de_n/dT \neq 0$ for $T \neq 0$, therefore $dS_c/de_n = 0$ is obtained as the condition which satisfies the extrema requirement. Thus

$$\frac{dS_c}{de_n} = \frac{C_\infty N_T}{2N_B} [-t_2 \exp(-e_n t_2) + t_1 \exp(-e_n t_1)] = 0$$

and solving for e_n^{max}

$$e_n^{max} = \frac{\ln(t_1/t_2)}{(t_1 - t_2)} \quad . \tag{17}$$

The relaxation time τ_n is usually used to describe the defect emission process ($\tau_n \equiv 1/e_n$). In this case $\tau_n^{max} = \dfrac{(t_1 - t_2)}{\ln(t_1/t_2)}$.

For the case of current transients, the observed signal, S_i, (in the case of low reverse bias leakage currents) is given by

$$S_i = i(t_1) = 1/2 \, AqWN_T e_n \exp(-e_n t_1) \quad .$$

The extrema condition $dS_i/de_n = 0$ yields

$$e_n^{max} = 1/t_1 \text{ or } \tau_n^{max} = t_1 \quad . \tag{18}$$

The extrema conditions for either capacitance or current measurements mean that for a given time of measurement the temperature at which the maximum signal occurs is related to a certain τ_{max}. A change in the measurement time results in a new τ_{max} and a shift in temperature for the peak signal. From Eq. (16) we find that

$$\tau_{max} T^2 = K \exp \left(+(E_C - E_T)/kT\right)$$

where $K = \dfrac{2.843 \times 10^{-22}}{\sigma_n}$. Thus an Arrhenius plot of $\ln(\tau_{max} T^2)$ vs. 1/T will have a slope of $(E_C - E_T)$ and the intercept can be used to calculate σ_n. Since the height of the peak is directly related to N_T, the following parameters can be obtained from DLTS and TCS: E_T, σ_n, and N_T for each defect level observed. A word of caution concerning the interpretation of E_T and σ_n. In the above analysis of the basic DLTS and TQS techniques, it has been assumed that σ_n does not depend on temperature. Unless the temperature dependence of σ_n has been determined by an independent method, it is best to consider both E_T and σ_n as signatures of deep levels rather than accurate measurements of the trap energy and the capture cross sections. A more detailed discussion of the various experimental techniques may be found in Ref. (6).

PAST APPLICATIONS OF DLTS TO NEUTRON IRRADIATED SILICON

This paper will not attempt to review all of the previous work (much of which is contained in the previous NTD Conference Proceedings[7],[8]). Instead an example which demonstrates the sensitivity of DLTS and the ability of DLTS to discriminate a large number of defect signatures will be presented. A series of anneals of neutron irradiated n- and p-type Czochralski Si are shown in Fig. 3. Seven distinct electron traps in n-type silicon are observed and fourteen hole traps in p-type. A unique signature has been obtained for all of these levels.[9] Clearly an abundant amount of information can be obtained through DLTS.

It is, at best, very difficult to obtain microscopic defect information from DLTS studies alone. Therefore, the application of DLTS becomes most useful when the DLTS results can be correlated with other techniques. Since there exists a large body of information concerning defects in Si (EPR, Raman scattering, photoluminescence, etc.), it has been possible to make a tentative association of nine DLTS levels with previously identified defects.[9] While the identification of a particular defect with a deep level observed in DLTS may be very useful, it may be even more useful when DLTS results can be correlated with device performance. An example of correlation of DLTS with lifetime in NTD-Si has been given elsehwere in these Proceedings.[10]

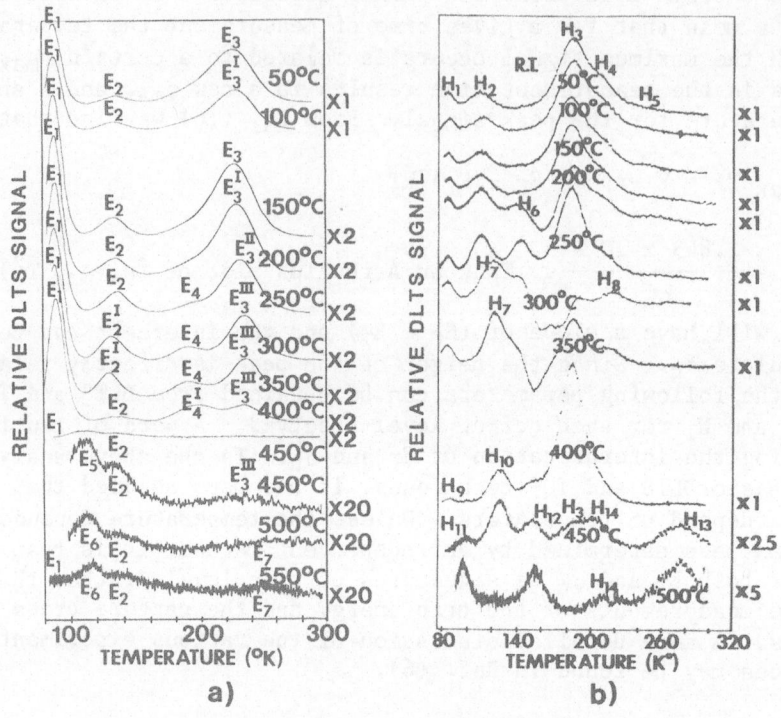

Fig. 3.   DLTS levels in neutron irradiated Si as a function of
          isochronal anneals.
          a) electron traps in n-type Si (100 Hz, ϕ_{th} =
              1.8 x 10^{15} n/cm²).
          b) hole traps in p-type Si (10 Hz, ϕ_{th} =
              1.8 x 10^{16} n/cm²).

TCS IN NEUTRON IRRADIATED SILICON

TCS has two advantages over the standard DLTS. Current measurements can be made more rapidly than capacitance. The increased
speed means that defect levels can be studied over an extended
range of emission rates. Secondly, even though in principle the
signal-to-noise ratio obtainable with capacitance measurements
should be similar to that of current measurements, it has been
found that the capacitance sensitivity is $N_T \approx 10^{-5} N_D$ while current sensitivity is $N_T \approx 10^{-7} N_D$.[11]

An Arrhenius plot of TCS data is shown in Fig. 4. The data
were obtained with a very simple apparatus in which the current was
integrated to eliminate the dependence of the signal amplitude on
emission rate.[12] The integration made it possible to maintain excellent sensitivity over the entire range of measurements. The

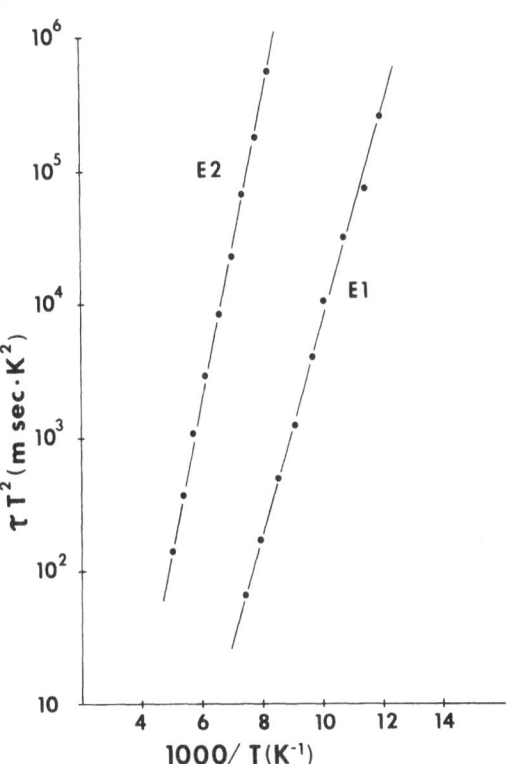

Fig. 4.   Arrhenius plot of TCS data for neutron irradiated Si.

data in Fig. 4 yields E1 = 0.15 eV, σ_1 = 2.3 x 10^{-15} cm^2 and E2 = 0.22 eV, σ_2 = 8.7 x 10^{-16} cm^2. The trap energies are in very good agreement with capacitance measurements (on the same samples and also in the literature[9]). The cross sections differ only slightly. The value obtained for the cross section is very sensitive to slight variations in the slope of the Arrhenius curve. Thus, since the TCS data covers a wider range of τ the TCS determination of σ_n is probably more precise.

The excellent signal-to-noise ratio of the TCS technique makes it possible to obtain a more detailed spatial defect profile than is possible with DLTS. The profile information is obtained in the following manner. If, instead of pulsing to zero volts, the diode is pulsed to some voltage V_1, such that the depletion width is W_1, the current will be

$$i_1(t) = \frac{Aqe_n}{W} \int_{W_1}^{W} (W - X)\, n_T(x)dx \quad . \tag{19}$$

W is the depletion width when the diode is reverse biased. The current can be obtained for a second width, W_2. The difference will be

$$\Delta i = i_2(t) - i_1(t) = \frac{Aqe_n}{W} \int_{W_2}^{W_1} (W - X)\, n_T(x)dx \quad . \tag{20}$$

If $|W_2 - W_1| \ll W$ then $n_T(x) \approx$ constant = $n_T(W_1)$ and

$$\Delta i = Aqe_n n_T(W_1)[W_1 - W_2 - \frac{1}{2W}({W_1}^2 - {W_2}^2)] \tag{21}$$

If $W_2 = W_1 + \Delta W$ where $\Delta W \ll W_1$ then

$$n_T(W_1) = \left(\frac{\Delta i}{\Delta W}\right) \frac{W}{Aqe_n(W - W_1)} \quad . \tag{22}$$

The results of profiling E1 and E2 are given in Fig. 5. One caveat is in order. The profiling analysis is based on the assumption of an abrupt junction. The profiling data in Fig. 5 are from a sample for which the assumption of an abrupt junction may not be valid. Thus, while it appears that E2 is constant and that E1 varies with depth, it is perhaps safer to state only that relative concentration of E1 to E2 varies with depth.

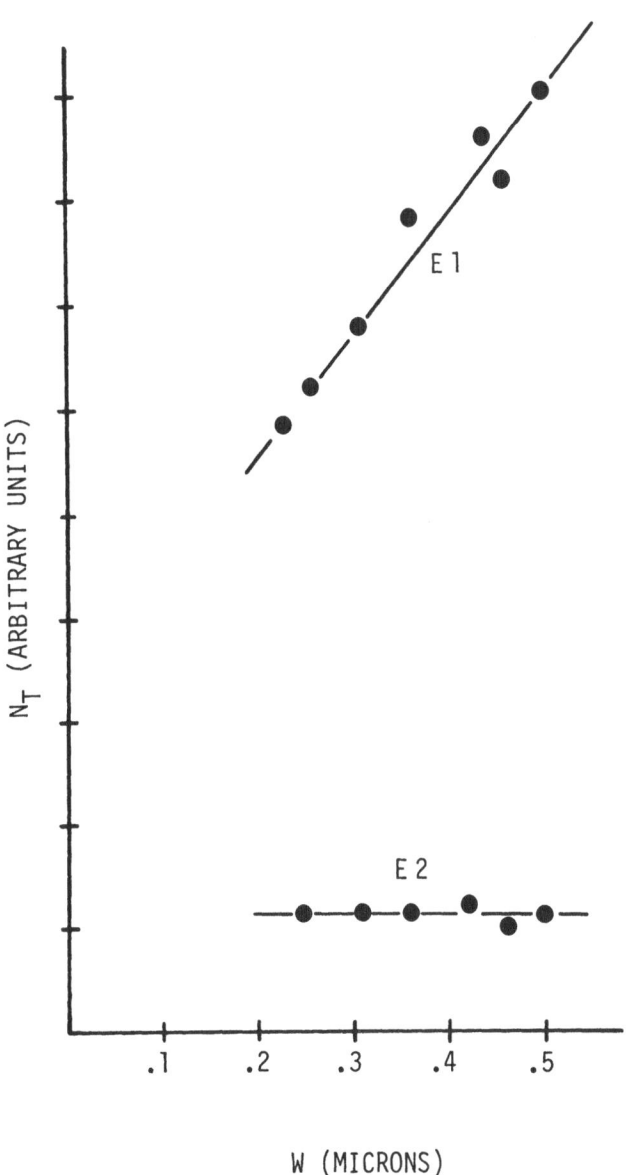

Fig. 5. Profile data for traps E1 and E2.

CONCLUSION

DLTS is a powerful "defect spectroscopic" technique. Over twenty uniquely identifiable deep levels have been observed in neutron irradiated Si. While DLTS (and TCS) may be somewhat limited in providing microscopic defect information, comparison of the wealth of information obtained from DLTS with that available from other techniques has led to a tentative association of nine of the DLTS levels with specific known defects in Si. The application of DLTS to device processing of NTD-Si has already been fruitful in identifying where processing problems exist.

While in principle TCS is simply a variation of the standard DLTS technique, TCS provides some distinct advantages over DLTS. The increased speed of current measurements has extended the range of emission rates over which defects may be characterized. The increased emission rate range makes possible the study of shallower defect levels as well as increasing the precision of the "signature" parameters E_T and σ.

TCS has also proved to be a very sensitive technique for the study of deep levels. Trap concentrations on the order of 10^{-7} times the doping level are observable with TCS. Such a high sensitivity is particularly valuable in the application to NTD processing where very small numbers of defects can cause a large reduction in minority carrier lifetimes. The sensitivity of TCS also makes possible detailed spatial profiling. The ability to profile is very useful in identifying those defects which may tend to migrate during processing and/or during device operation. The latter mechanism may give rise to premature device failure.

In general, both DLTS and TCS have proved to be very useful in the study of deep levels in semiconductors. Continued application of both techniques to NTD-Si should be very useful in all aspects of materials processing, from the irradiation to the final device product.

REFERENCES

1) C. T. Sah, L. Forbes, L. L. Rosier and A. F. Tasch, Jr., Solid-State Electronics 13, 759 (1970).
2) C. T. Sah, Semiconductor Silicon, 1977, ed. by R. Huff (Electrochem. Soc., Princeton, NJ, 1977), p. 868.
3) C. T. Sah, Solid-State Electronics 19, 975 (1976).
4) G. L. Miller, D. V. Lang and L. C. Kimerling, Ann. Rev. Mater. Sci. 377, (1977).
5) D. V. Lang, J. Appl. Phys. 45, 3023 (1974).
6) J. W. Farmer and J. M. Meese, J. Nuc. Mater. 108 + 109, 700 (1982).

7) Neutron Transmutation Doping in Semiconductors, ed. by J. M.
    Meese (Plenum Press, New York, 1979).

8) Neutron-Transmutation-Doped Silicon, ed. by J. Guldberg
    (Plenum Press, New York, 1981).

9) J. M. Meese, M. Chandrasekhar, D. L. Cowan, S. L. Chang, H.
    Yousif, H. R. Chandrasekhar and P. McGrail, in: Neutron-
    Transmutation-Doped Silicon, ed. by J. Guldberg (Plenum Press,
    New York, 1918), p. 101.

10) D. E. Crees and P. D. Taylor, "Process Induced Recombination
    Centres in NTD Silicon and Their Influences on HVDC
    Thyristors," this conference.

11) J. A. Borsuk and R. M. Swanson, IEEE Trans. Elec. Dev. ED-27,
    2217 (1980).

12) J. W. Farmer, C. D. Lamp and J. M. Meese, Appl. Phys. Lett.
    41, 1063 (1982).

CALIBRATION OF THE PHOTOLUMINESCENCE TECHNIQUE FOR

DETERMINATION OF PHOSPHORUS IN SILICON BY NEUTRON

TRANSMUTATION DOPING

Bobbie D. Stone[1], Aliene D. Henry[1], Paul L. Clem, Jr.[1],
Larry W. Shive[1] and Steve L. Gunn[2]

1. Monsanto Electronics Division, St. Peters, Missouri
2. University of Missouri Research Reactor, Columbia, Mo.

SUMMARY

A rapid accurate method for characterizing polycrystal silicon
for residual electrically-active impurities has been developed based
on the simultaneous determination of boron, phosphorus, arsenic and
aluminum in the 10^{-10} atoms/atom range in silicon monocrystal by
photoluminescence. The method was originally calibrated on the basis
of resistivity, but this leads to a relatively large uncertainty,
especially for donor impurities. This paper describes the recalibration of the method for phosphorus by NTD using a carefully controlled
neutron fluence. The agreement between the three methods of measuring impurity content are presented and the statistical reproducibility of the method is discussed.

INTRODUCTION

Since the simultaneous determination of boron and phosphorus in
silicon crystals by photoluminescence was reported by Tajima[1], the
potential application of this method to the characterization of
electronic-grade polysilicon has been apparent. The standard method
of analyzing for phosphorus and boron in silicon consists of making
different numbers of vacuum zone refining passes (typically one and
six) on separate portions of a single rod and preparing single
crystal sections for analysis. Since a known fraction of the phosphorus is evaporated per zone pass at a given zone travel rate,
measurement of electrical resistivity on the two crystals allows the
determination of the absolute levels of boron and phosphorus by the

241

solution of simultaneous equations. In addition to being time con-
suming and expensive, however, this classical method has inaccuracies
due to the inherent difficulty in controlling the parameters affecting phosphorus evaporation rates and has no provision for dealing
with other donors and acceptors such as arsenic and aluminum. By
contrast, the photoluminescence method requires only a very small
sample of monocrystal easily obtainable by a single zone pass in gas
and gives independent, unambiguous measurements of arsenic and
aluminum as well as phosphorus and boron. During the past two years,
we have actively pursued a program for establishing a polysilicon
evaluation procedure based on photoluminescence and in the process
have correlated resistivity with net impurity content determined by
this method for over 2000 samples.

Tajima and co-workers[2,3] have shown also that photoluminescence
can be used to measure phosphorus introduced in silicon by Neutron
Transmutation Doping (NTD). This paper will describe the calibration of our system in absolute terms using resistivity of seven-pass
vacuum refined crystal for boron and NTD of phosphorus- and boron-
free crystals for phosphorus.

The analysis procedure for polycrystal is shown schematically
in Figure 1. A small core is cut from a crack-free section of the
polycrystal sample, etched and converted to a monocrystal by a single
float-zone pass in argon. A 0.5 mm thick sample is cut from the rod,
bright-etched and analyzed by the photoluminescence (PL) technique.
The procedure was described in detail at the Montreal meeting of the
Electrochemical Society in May 1984[4].

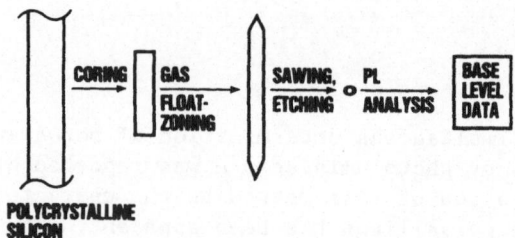

Figure 1. A Schematic Diagram of the Procedure
for Base Level Analysis of Polycrystalline Silicon.

EXPERIMENTAL

The samples used in the experiments were cut from a rod approx-imately 19 mm in diameter that had been subjected to seven passes in high vacuum. (Graciously donated by Hemlock Semiconductor Corpora-tion.) One set consisted of slugs approximately 15 mm thick cut from rods of varying boron content and these were used to calibrate the PL instrument for boron. The PL analyses showed these to contain a maximum of 0.01 ppba phosphorus. The other set consisted of two rods about 120 mm long with a 4-point resistivity in excess of 4000 ohm-cm "P" type. Samples of the tang end of each of these were analyzed by PL and found to be < 0.01 ppba for both phosphorus and boron. The rods were cut into slugs approximately 15 mm thick for the irradiation experiments.

The photoluminescence apparatus used is shown schematically in Figure 2. Light from an argon laser (Control Laser Corporation, Model 551A, 0.2 watt) is conducted through a series of mirrors and lenses and caused to impinge on a bright-etched surface on the silicon sample immersed in liquid helium. The spectrum of infrared radiation luminesced by the sample is resolved by a monochromator, (Spex Industries, Inc., Model 1870 $\frac{1}{2}$ meter spectrograph with Model VPM-159) and recorded on a multispeed recorder (Fisher Scientific Company, Model RF 110137). The resultant spectrum contains a series of peaks of light intensity at specific wavelengths. Each of the peaks except the intrinsic silicon peak is associated with specific impurities in the silicon. It is the ratio of the intensity of each impurity peak to that of the intrinsic peak that is the basis of the quantitative calculation.

The initial calibration of the method for boron was established by measurement of silicon that had several vacuum passes to remove essentially all the phosphorus and the boron level was determined by resistivity. This empirical relationship for boron was found to be:

$$\log [B] = 1.16 \log \frac{I_B}{I_0} + 12.84$$

where [B] is the boron concentration (in atoms/cm^3) and I_B and I_0 are the intensities of the boron and intrinsic signals, respectively. For the original phosphorus calibration, no boron-free samples were available, and it was necessary to deduce the phosphorus concentra-tion from the resistivity which is due to the net difference between phosphorus and boron concentrations. For example, with more phos-phorus than boron, this relationship would be:

$$\text{Resistivity (ohm-cm)} = \frac{5.4 \times 10^{15}}{([P] - [B])} \text{ (atoms/cc)}$$

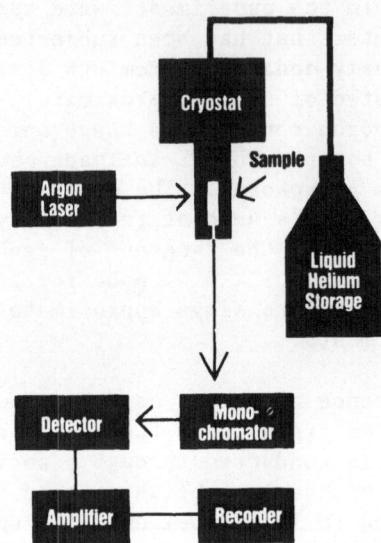

Figure 2. The Apparatus for Polysilicon Assay by Photoluminescence.

Although this gives an unambiguous value for phosphorus in the
sample, it involves both the errors associated with measurement of
resistivity and with the PL measurement of boron. In some cases
levels of phosphorus deduced by this method were based on a
relatively small difference between two large numbers. The NTD
approach, which gives a totally independent determination of
phosphorus, is ideally suited to circumvent these difficulties.

Pairs of samples (one sample from each rod) were encapsulated
in a 5" long aluminum can with flux wires (NBS 0.116% cobalt in
aluminum) placed at each end and between the samples. These sample
capsules were then irradiated in the I-2 position in the graphite
reflector in the 10 megawatt light water Research Reactor at the
University of Missouri at Columbia, Mo. The flux field was measured
at $4.6 \pm 0.10 \times 10^{13}$ neutrons/cm^2/sec, and the thermal/fast ratio
was approximately 27:1. The various samples were scheduled to yield
0.5, 1,0, 1.5, 2.0, 2.5 and 3.0 ppba phosphorus based on the well-
established relationship of 1 ppba $\sim 2.98 \times 10^{17}$ neutrons/cm^2. The
fluence was measured by the control system comprised of rhodium wire
self-powered neutron detectors and analogue current integrators[4] and
by counting the NBS flux wires. (The mean standard deviation for
counting was 2.5%.)

In the standard process for NTD float-zone silicon the annealing procedure for removing residual irradiation damage was developed to give (1) stable resistivity or carrier concentration; and (2) maximum minority carrier lifetime. This process consists of passing the cleaned irradiated crystal through a temperature zone of 750°-825°C for a period of about 75 minutes and cooling at a rate of < 2°/min to about 400°C. One set of samples (the "B" group) was annealed by this procedure, and after lapping off 0.1 mm, ten resistivity measurements were made on the lapped surface. After bright etching in an $HF-HNO_3-CH_3CO_2H$ mixutre, the samples were then annealed at 1150°C in a wet oxygen atmosphere. The procedure in this case was to move the samples into the furnace at 800°C at 3"/min, "ramp" the furnace to 1150°C, leave at this temperature 3 hours, "ramp" down to 800°C and remove at a pull rate of 3"/min. After stripping the oxide and lapping about 0.1 mm off the surface, a series of 10 resistivity readings were again taken on the lapped surface. The other set of samples (designated as "A") were subjected to the same treatments, but the resistivity measurement after the first anneal at 750°-825°C was inadvertently omitted.

These annealing experiments were performed in an effort to establish the stable resistivity level to compare with PL and NTD fluence values. Although long experience with the 750°-825°C annealing process had indicated a stable resistivity, the results of Tajima and Yusa raised some doubts. Although our primary purpose is to relate NTD fluence data with PL results, the correlation with resistivity is a very important side issue.

Table I. Comparison of Boron Level Determined by Photoluminescence with Those Based on Resistivity in Phosphorus-Free Samples. (Values in ppba)

N_B (Resistivity)*	N_B(PL)			
	1st DET'N	2nd DET'N	$\bar{X}$	σ
.019	.01	.01	.01	0
.034	.02	.02	.02	0
.060	.04	.04	.04	0
.095	.07	.08	.075	.007
.120	.06	.08	.07	.014
.220	.19	.22	.205	.021
.330	.33	.31	.32	.014
.590	.64	.55	.595	.064

*Based on Relationship $N_B = \dfrac{273}{\rho \text{ (ohm-cm)}}$

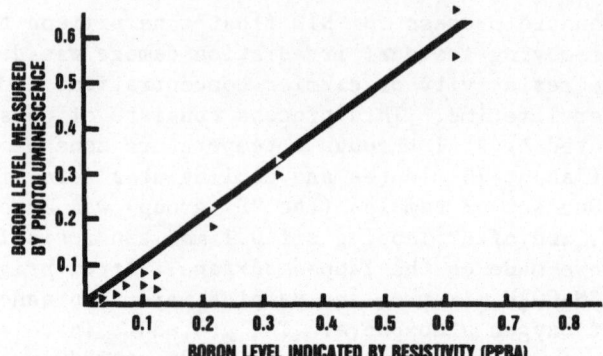

Figure 3.   Comparison of Boron Level Measured by Photoluminescence
            with That Deduced from Resistivity in Donor-Free Samples.

Resistivity measurements were made with a 4-point probe with
0.025" probe spacing at a constant current of 1 ma up to 80 ohm-cm
and 0.1 ma above 80 ohm-cm. Temperature was controlled at 23 ± 2°C
and relative humidity at 60 ± 2%. In each case a 0.5 mm thick
sample was cut from the surface where the resistivity measurements
were made, the sample bright etched and analyzed by the photoluminescence technique.

RESULTS

The results of PL analyses for the boron-doped samples and
their comparison with the resistivity determinations are shown in
Table I. In all cases the reproducibility is very good with the
mean standard deviation being only 7.4%. Agreement between the two
methods is excellent, especially above about 0.1 ppba. The calibration was carried out only to 0.6 ppba, but this covers the region of
interest in present day semiconductor-grade polysilicon. The data
are presented graphically in Figure 3.

The phosphorus values determined by PL are compared to those
deduced from neutron fluence in Table II. The fluence data are
based on the mean value of the three flux-wire measurements. The
plot of these data shown in Figure 4 reveals that the original
calibration based on resistivity was in error above about 1 ppba
with the PL values being about 20% high at 3 ppba. Again the area
of major interest for semiconductor-grade polysilicon is below 1.0
ppba phosphorus, so no serious error was involved in earlier data.
A least-squares line drawn through the points displayed in Figure 4
was used to give a new phosphorus relationship with peak height.

Table II. Comparison of Phosphorus Levels Determined by
Photoluminescence with Those Predicted by Neutron
Fluence. (Values in ppba)

SAMPLE NO.	N_P^* (FLUENCE)	N_P(PL)			
		1st DET'N	2nd DET'N	$\bar{X}$	σ
1A	0.52	0.58	0.47	0.53	0.08
1B	0.52	0.60	0.58	0.59	0.01
2A	1.01	1.09	1.04	1.07	0.04
2B	1.01	1.23	1.18	1.21	0.04
3A	1.59	1.29	1.63	1.46	0.24
3B	1.59	2.85	2.14	2.49	0.49
4A	2.15	2.62	2.42	2.52	0.14
4B	2.15	2.77	2.52	2.65	0.18
5A	2.55	3.17	2.94	3.06	1.16
5B	2.55	3.06	3.74	3.40	0.47
6A	3.12	3.56	4.16	3.86	0.42
6B	3.12	3.87	4.16	4.02	0.21
C-1	1.05	1.45	1.57	1.51	0.08
D-2	1.05	1.42	1.46	1.44	0.03

* Calculated from relationship 1 ppba $\sim$ 2.98 x 10^{17} neutrons/cm^2.

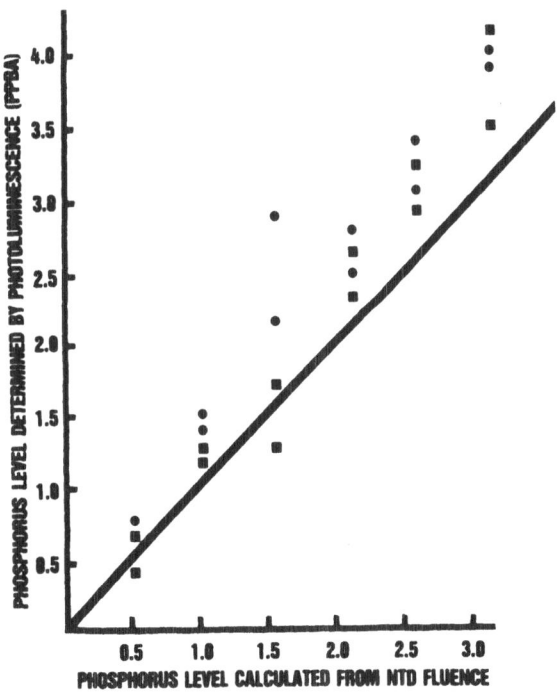

Figure 4. Comparison of Phosphorus Levels Determined by PL in
Boron-Free Silicon with Those Deduced from NTD Fluence.

The relationship between resistivity and annealing temperature appears to be more complicated and, although not nearly as extensive, the results of our experiments did not agree with those of Tajima and Yusa[3]. Resistivity values calculated from the results of the PL analyses and the measured NTD fluence are compared with measured resistivities in Table III. Comparing the last two columns, it is apparent that the measured values for the "B" samples agree much better with the NTD values than do those for the "A" samples. At the stage described in this table, the "B" samples had been annealed by only the standard NTD float zone process involving holding the sample in the 750°-825°C range for about 1.25 hours, while the "A" samples had been subjected to an additional three-hour anneal at 1150°C before the resistivity was measured. On the other hand the authors cited found that annealing at 800°C for longer than 10-15 minutes gave resistivities that decreased with time with the rate of decrease increasing as the thermal/fast neutron ratio decreased. Either very long annealing times (30 hours) at 800°C or short anneals at 1200°C were necessary to regain the target resistivity level. In our case the "A" samples annealed at 1150°C gave resistivity values about 15% below the target values, indicating spurious generation of carriers at the higher annealing temperatures. Moreover, when the "B" samples were re-annealed at 1150°C for three hours, the resistivity level fell to a point almost as low as for the "A" samples (see Table IV).

Table III. Comparison of Resistivities Predicted by Photo-luminescence with Those Predicted by NTD Fluence and with Measured Values. (Values in ohm-cm)

SAMPLE NO.	PL VALUES			NTD FLUENCE	
	$\bar{X}N_B$	$\bar{X}N_P$	ρCALC'D	ρCALC'D[1]	ρ(MEASURED)[2]
1A	.04	.52	200	184	155 ± 2
1B	.04	.59	175	184	178 ± 2
2A	.06	1.6	96	95	78 ± 1.1
2B	.08	1.21	85	95	91 ± 0.9
3A	.07	1.46	71	60.4	67 ± 1.4
3B	.14	2.49	41	60.4	61 ± 0.8
4A	.13	2.52	40.2	44.6	38.7 ± 0.5
4B	.13	2.65	38.1	44.6	45.4 ± 0.5
5A	.13	3.05	32.9	37.7	31.3 ± 0.3
5B	.16	3.40	29.6	37.7	37.0 ± 0.3
6A	.18	3.86	26.1	30.8	26.3 ± 0.4
6B	.19	4.02	25.1	30.8	30.8 ± 0.4
C-2	.47	1.57	92.3	128	116 ± 4
D-1	.60	1.44	114	209	165 ± 4

[1]-Calculated assuming original N_B and N_P of 0.01 ppba each.
[2]-Average of 10 separate readings ± one standard deviation.

Table IV. Comparison of Resistivity Values Measured on NTD
           Samples Annealed at Different Temperatures.
           (Values in ohm-cm)

SAMPLE NO.	"B" SAMPLES ANNEALED AT 750°-825°C	1150°C	"A" SAMPLES ANNEALED AT 750°-825°C#	1150°C	CALC'D FROM NTD FLUENCE
1	178.4	161.0	--	155.0	184.0
2	91.0	81.1	--	77.6	95.0
3	61.2	52.1	--	66.9*	60.4
4	45.2	40.0	--	38.7	44.6
5	37.0	32.6	--	31.3	37.7
6	30.8	27.2	--	26.3	30.8

\# The "A" samples were not measured after this anneal.
* Annealed twice at 1150°C for three hours.

The variability of the various measurements are summarized
in Table V. Except for sample "3B", which gave an anomalously high
PL value for phosphorus, the standard deviation for the total of 4
determinations at each level (two on each sample) was < 11%. The
variation in the resistivity values was almost as great, but this
includes variations due to different annealing treatments. Note
that the resistivity variability within a sample was < 3% in all but
one case.

Table V. Standard Deviations for Values of Phosphorus Determined
by Photoluminescence, Resistivity Measurements, and NTD Fluence.

SAMPLE NO.	PL VALUES STD. DEV'S (%)			RESISTIVITY VALUES STD. DEV'S (%)			NTD FLUENCE STD. DEV'S (%)	
	Within Sample	Between Samples	Overall	Within Sample	Between Samples	Overall	Within Wires	Overall
1A&B	8.5	8.0	11	1.2	9.8	7.2	4.9	4.6
2A&B	3.1	8.7	7.6	1.2	10.9	8.2	3.1	2.6
3A&B	18.2	36	34.3	1.7	6.6	4.6	0.3	3.0
4A&B	6.2	3.6	5.8	6.5	11.3	8.2	2.2	4.0
5A&B	9.5	7.4	10.8	0.9	11.8	8.6	1.8	1.8
6A&B	8.1	2.9	7.3	1.4	11.1	8.3	1.8	1.8
1C&2D	3.7	3.4	4.4	2.9	-	-	4.5	4.4

CONCLUSIONS

The original calibration of the PL method for boron in the
absence of phosphorus proved to be correct and in the range from
0.1 to 0.6 ppba, the method appears to be extremely accurate. After
introduction of phosphorus by NTD, however, the detected level of

boron seemed to be unrealistically high with the apparent level increasing with increased phosphorus level (see Table III). This was found to be due to a simplifying assumption in the original mathematics, and was corrected along with the phosphorus calibration.

The earlier calibration for phosphorus was found to be in error with PL being on the high side by about 20% at the 3 ppba level. The equation expressing the relationship between phosphorus concentration and peak height of the phosphorus peak was changed from

$$\log [P] = 1.261 \log 2.3 \frac{Pi}{I} + 12.7 \quad \text{to}$$

$$\log [P] = 1.155 \log 1.88 \frac{Pi}{I} + 12.84$$

Agreement between the phosphorus levels measured by PL, by NTD fluence and resistivity indicate that a stable resistivity is obtained by a 1.25-hour anneal in the 750°-825°C range. Annealing at 1150°C appears to introduce spurious levels of donor carriers, in apparent disagreement with the results of Tajima and Yusa.

With the exception of one sample the reproducibility of the PL results is such as to have a standard deviation of < 11%. Since the values given here were used to recalibrate the PL method for phosphorus, estimation of the absolute accuracy of the method was difficult. After the recalibration and correction of the boron equations, it appears to be considerably better than the ± 40% standard deviation that had been deduced from resistivity considerations alone.

REFERENCES

1.   M. Tajima, Determination of Boron and Phosphorus Concentration in Silicon by Photoluminescence Analysis, Appl. Phys. Letters 32:719 (1978).
2.   M. Tajima, Characterization of Neutron-Transmutation Doping in Silicon by Photoluminescence Technique, Appl. Phys. Letters 35:242 (1979).
3.   M. Tajima and A. Yusa, Characterization of NTD Silicon Crystals by the Photoluminescence Technique, in: "Neutron Transmutation-Doped Silicon". J. Guldberg, ed., Plenum Press, New York and London, (1981)
4.   L. W. Shive, "Photoluminescence Analysis -- Tool for Poly-crystal line Silicon Quality Evaluation", Spring 1982. Meeting of the Electrochemical Society, Montreal, Quebec, Canada, May 9-14, 1982. Paper No. 190.
5.   S. L. Gunn, J. M. Meese and D. M. Alger, High Precision Irradiation Techniques for NTD Silicon at the University of Missouri Research Reactor, in: "Neutron Transmutation Doping in Semiconductors", J. M. Meese, ed., Plenum Press, New York and London (1979).

SWIRLS IN NEUTRON-TRANSMUTATION DOPED

FLOAT-ZONED SILICON

Horst G. Kramer

Electronics Division
Monsanto Company
St. Peters, Missouri   63376

ABSTRACT

Today, dislocation-free float-zoned ingots are routinely grown free of swirls. However, after neutron-transmutation doping and after annealing to restore electrical properties, these ingots may show swirls after preferential etching. Statistical evidence indicates that, if both oxygen and carbon are present in concentrations less than 1 ppma (5×10^{16} atoms/cm^3), then it is oxygen that causes swirls to form. The lower threshold of oxygen for swirl formation appears to be 0.15 ppma. This hypothesis has been experimentally verified by adding small amounts of oxygen during regrowth to previously swirl-free neutron-transmutation doped ingots. A second round of irradiation and annealing then produced swirls in these ingots. The paper presents evidence that neither the zone refiner nor the zone-refining process induced swirls, so that the most likely source of the oxygen is the polycrystalline silicon itself.

INTRODUCTION

Swirls in dislocation-free crystals of silicon, grown by both the float-zone and the Czochralski method, have been the subject of intensive study for many years. These swirls are revealed by preferential chemical etches such as the Sirtl or the Wright etch and appear as spiral defect patterns on ingot surfaces cut at right angles to the growth axis. They also appear as alternating lines of high and low defect density in the shape of the solid-liquid growth interface on ingot surfaces cut parallel to the crystal growth axis. These swirl patterns can be observed with the unaided eye, and their geometric patterns make it obvious that

251

they were created during crystal growth.

When swirls are observed under the microscope, they can be seen to consist of regions of high density of etch pits. (Figures 1 and 2). Two types of defects appear, named A- and B-type by deKock[1] as early as 1970. Work by that author and co-workers[2] and by Föll and co-workers[3] has led to an understanding of the nature of these two defects. B-type defects are small, three-dimensional clusters of silicon interstitials, possibly containing carbon or oxygen atoms, formed during cooling of the crystal. These B-type defects will continue to grow upon further cooling, and, if they reach a critical size, will become unstable and collapse, forming faulted dislocation loops. Perfect dislocation loops are formed via a defaulting reaction, requiring the generation of Shockley partials. These are the A-type defects.

In 1973, deKock, Roksnoer and Boonen[4] showed that, in float-zoned ingots, the formation of both A- and B-type defects can be suppressed if the growth rate is above 5 mm/min. In 1977, Föll, Gösele, and Kolbesen[5] showed that carbon-rich silicon exhibits B-type defects at growth rates above 5 mm/min, but that at carbon concentrations of .1 ppma or less no B-type pits are formed. They came to the conclusion that the oxygen content of their crystals (ranging from .01 - .06 ppma) had no influence on B-pit formation. As a consequence of this work, today's float-zoned silicon doped by means other than neutron transmutation, such as vapor-phase or slim-rod doping, is free of swirls, although it will still contain B-type defects at low densities. In Czochralski ingots, with their higher oxygen and carbon contents, elimination of both A- and B-type defects has not been as rapid, and work continues. However, in Czochralski ingots the correlation between oxygen content (ranging from 20 to 40 ppma) and swirls is well documented.

The availability of neutron-transmutation doped (NTD) float-zoned silicon in commercial quantities from 1975 on led to its use in many applications where uniform resistivity across a wafer enhances the performance of the device built from the wafer. However, some such devices (Charge-coupled device (CCD) imagers, for example) are also extremely sensitive to microdefects in the silicon, particularly if the defects are incorporated inhomogeneously[6]. Attempts to use NTD wafers for CCD imagers showed that at least some of the wafers again exhibited the familiar swirl pattern.

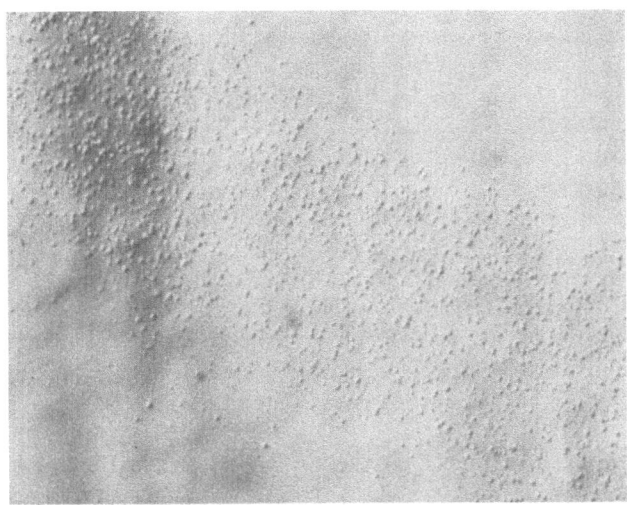

Figure 1. B-type Pits Forming a Swirl Pattern

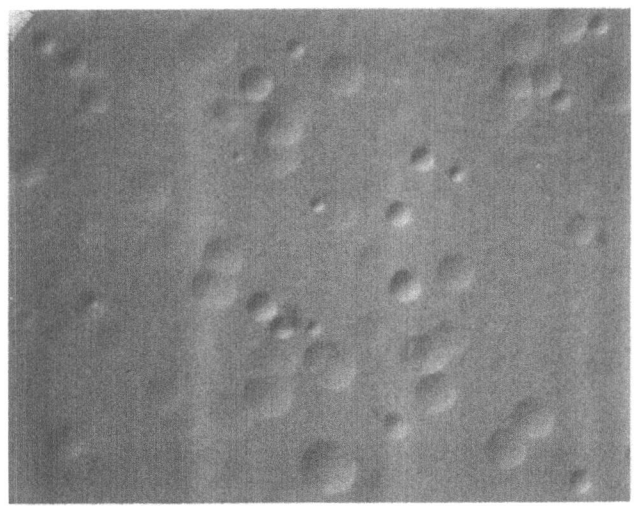

Figure 2. The Region Between Swirled and Swirl-
free Areas, at Larger Magnification

THE INITIAL SURVEY

Recently, the author undertook a survey of float-zoned silicon produced by various commercial, competitive companies. Where
only polycrystalline rods were available from a producer, they were
turned into dislocation-free ingots in Monsanto zone refiners.
From each ingot, four two-inch long sections were cut from the end
of the ingot farthest away from the seed. The four sections from
each ingot were processed as follows:

> Section 1 was left as grown, without any additional proces-
>   sing.
> Section 2 was annealed in a heat-treatment cycle used for
>   NTD material to restore resistivity and lifetime.
> Section 3 was neutron-transmutation doped, but not annealed.
> Section 4 was neutron-transmutation doped, and then annealed
>   in the same was as Section 2.

The neutron dose for Sections 3 and 4 was equivalent to adding
2 ppba of phosphorus. Irradiation was done at the Missouri
University Research Reactor at Columbia, Missouri. Each section
was then evaluated for oxygen and carbon content by Fourier-
transform infrared spectroscopy (FTIR) and for swirl formation by
Sirtl etching. To establish the precision of the FTIR method at
the predictably low levels of oxygen and carbon encountered,
repeat readings on the same wafers were taken and compared to
repeat readings of adjacent wafers of the same ingot section.
The results of the survey were as follows:

> · All as-grown and as-grown and annealed ingot
>   sections were free of swirl, regardless of
>   oxygen or carbon content.
>
> · Similarly, ingot sections having undergone neutron
>   irradiation only were free of swirls.
>
> · Swirls appeared only on wafers that had been both
>   irradiated and annealed.
>
> · Some of the irradiated and annealed wafers did not
>   show swirl patterns.

Table I shows the results of the carbon and oxygen analyses
performed on all wafers. The numbers in the table are average
concentrations for as-grown and as-grown and annealed ingot sections. The indication of swirl is for irradiated and annealed
sections. (FTIR measurements of the irradiated-only wafers showed
no measureable oxygen or carbon content, because of the lattice
damage introduced by NTD.[7]) Even a cursory look at the table
shows that the only correlation is between oxygen content and

swirls. At levels of about 0.15 ppma of oxygen or higher, swirls
begin to appear. Moreover, those wafers exhibiting swirls show a
greater reduction of the oxygen content from the as-grown to the
irradiated and annealed sections. This can be explained by
postulating the precipitation of oxygen atoms during annealing on
lattice damage sites introduced by the irradiation. This data is
presented in Table 2. A statistical analysis of the data presented
in Table 2 shows that the difference in average oxygen concentration of ingots with swirls and ingots without swirls is highly significant; the reduction in oxygen concentration after irradiation
and annealing of ingots with swirls is significant. The carbon
differences are not significant.

MORE DATA

After the results of the previous section became available,
a number of float-zoned ingots were analyzed for oxygen and carbon
before being irradiated, and the incidence of swirls was determined
on the same ingots after irradiation and annealing. The results
for carbon were similar to those presented in the previous section,
and showed no correlation. The results for oxygen are given in
Table 3 and show a highly significant dependence of swirl appearance as a function of the oxygen level. The oxygen threshold level
of 0.15 ppma, indicated by the data of Table 1, holds up rather
well.

ESTABLISHING A CAUSE-AND-EFFECT RELATIONSHIP

The evidence presented so far is of a statistical nature.
However, methods of demonstrating a direct cause-and-effect relationship are readily available. By making an additonal zone pass
over previously swirl-free crystals and adding oxygen during
growth, the oxygen content of the crystals can be increased.
Conversely, by making an additional vacuum pass over crystals
exhibiting swirls, their oxygen content can be reduced, since it is
well known that oxygen readily leaves a silicon melt in vacuum.
The effect of these additional passes on carbon is a reduction in
its content, since the segregation coefficient of carbon in silicon
is less than unity. Moreover, by making "neutral" passes which
neither add nor subtract oxygen from previously swirl-free ingots,
any contribution of the zone-refiner and of the zone-refining
process to swirl formation can be determined. All three of these
experimental methods were followed, but results from only two of
them are available. In the third method, polycrystalline material
for the combination consisting of a first vacuum pass followed by
a second argon pass was selected from a group of materials likely
to have an oxygen concentration higher than the postulated threshold of 0.15 ppma, and thus likely to produce swirls in the section
of ingot that had been subjected to only passes in argon.
However, the measurements taken after completion of the experi-

TABLE 1. SWIRLS VS. CARBON AND OXYGEN

Silicon from Sample Number	Refined by	Carbon Conc., PPma	Oxygen* Conc., PPma	Swirls
A/2	Monsanto	<.10	.25	Yes
B/1	Monsanto	.22	.22	Yes
A/1	Monsanto	.15	.20	Yes
E	Monsanto	.13	.18	Yes
D	Monsanto	.41	.14	Yes
F	F	.12	.12	Yes
B/2	Monsanto	.16	.12	No
C/1	C	.50	.06	No
C/2	C	.22	.06	No

* Oxygen concentrations were calculated according to ASTM F 121-76 with $O_i = 4.81 \times 10^{17} \alpha$ atoms/cm^3 or $O_i = 9.63\alpha$ ppma where α is the net absorptivity due to interstitial oxygen.

TABLE 2.   THE STATISTICAL EVIDENCE FOR OXYGEN
CAUSING SWIRLS IN NTD FLOAT-ZONED INGOTS

	Oxygen Conc., ppma	Carbon Conc., ppma
Precision of FTIR Measurements, 3-sigma Limits	$\pm.075$	$\pm.156$
Average Concentrations Before NTD		
of Ingots With Swirl	.19	.16
of Ingots Without Swirls	.10	.14
Average Reduction in Concentration After NTD and Annealing		
of Ingots With Swirls	.027	$-.07$
of Ingots Without Swirls	.009	.02

TABLE 3.   PRESENT STATE OF CORRELATION

The Numbers in the Body of the Table Show
the Number of Ingots With And Without Swirls

	Swirls	No Swirls
Oxygen Content <.15 ppma	3	29
Oxygen Content $\geq$.15 ppma	10	0

mental work showed that the oxygen content was below the detection
limit in both portions of the ingot, and neither portion exhibited
swirls after preferential etching.

The results of the other two methods are shown in Table 4.
Carbon and oxygen content and presence or absence of swirls are
shown for ingot sections before and after intentional doping with
oxygen. Ingots 1 through 4 represent the "neutral" passes, and
were not doped intentionally. The results of the first four ingots
of Table 4 indicate that the zone refiner or the zone-refining
process did not induce swirls in ingots that had been swirl-free
before. Of the remaining five ingots, two had swirls before
doping passes were made and despite the fact that the oxygen was
below 0.15 ppma. The other three ingots had no swirls and an
oxygen content below 0.15 ppma before doping passes were made, and
had swirls after various amounts of oxygen were added in the
doping passes.

CONCLUSIONS

The work presented in this paper has shown that neutron-
transmutation doped float-zoned silicon may exhibit swirl patterns
despite the fact that the as-grown ingot before irradiation and
annealing did not exhibit swirl patterns. It also has shown that
the presence of irradiation-induced crystal damage is necessary,
but not sufficient, to cause swirls. The fact that an annealing
step is required to produce swirls indicates that an impurity
precipitates on the damage sites. The results presented here
strongly suggest that this impurity is oxygen rather than carbon,
even though both are present in concentrations less than 1 ppma.

The oxygen very likely comes from the polycrystalline silicon
used to produce the dislocation-free ingot, since the zone refiner
did not add significant amounts of oxygen. The zone-refining
process does enter into the formation of swirls, since it distrib-
utes the oxygen (or any other impurity) in the typical helical
fashion throughout the ingot during growth.

The answer to the problem of swirls in NTD float-zoned ingots
hence appears to be the control of the oxygen content, either by
producing polycrystalline silicon with an oxygen content below the
detection of infra-red spectroscopy, or by making vacuum purifi-
cation passes before growing the final crystal. Although no
evidence has been presented for the effectiveness of such vacuum
passes, it is still felt to be the appropriate alternative for
removing oxygen and suppressing swirls.

TABLE 4. EFFECT OF OXYGEN CONCENTRATION CHANGES ON SWIRL FORMATION

Ingot Number	Before Doping Passes			After Neutral Or Doping Passes			
	Oxygen Conc., ppma	Carbon Conc., ppma	Swirls	Nature of Pass	Oxygen Conc., ppma	Carbon Conc., ppma	Swirls
1-4	<.10	.09	No	Neutral	.12	<.10	No
5	<.10	.10	No	Doping	1.03	.36	Yes
6	.14	.17	Yes	Doping	1.80	.43	Yes
7	.12	<.10	No	Doping	1.96	.08	Yes
8	.10	.40	No	Doping	2.81	<.10	Yes
9	.09	<.10	Yes	Doping	3.04	.12	Yes

REFERENCE

1.   A.J.R. de Kock, Applied Physics Letters 16, 100 (1970).

2.   A.J.R. deKock, Point Defect Condensation in Dislocation-free
      Silicon Crystals, in Semiconductor Silicon 1977, p.p. 508,
      H. R. Huff and E. Sirtl, editors, The Electrochemical Society,
      Princeton, N.J.

3.   H. Föll, H. O. Kolbesen, and W. Frank, Phys. Stat. Sol.
      (a), 1975, 29:83.

4.   A.J.R. de Kock, P. J. Roksnoer, and P.G.T. Boonen,
      Microdefects in Swirl-Free Silicon Crystals, in Semiconductor
      Silicon 1973, pp. 83, H. R. Huff and R. R. Burgess, editors,
      The Electrochemical Society, Princeton, N.J.

5.   H. Föll, U. Gösele, and B. O. Kolbesen, Swirl Defects in
      Silicon, in ref. 2, page 565.

6.   L. Jastrzebski, P. A. Levine, W. A. Fisher, A. D. Cope,
      E. D. Savoye and W. N. Henry, Cosmetic Defects in CCD
      Imagers, Journal Electrochemical Society, Vol. 128, No. 4,
      April 1981, pp. 885.

7.   R. C. Newman, Impurity Interactions with Structural Defects
      in Irradiated Silicon, in "Neutron-Transmutation Doped
      Silicon", Jens Guldberg, editor, Plenum Press, New York (1981).

ACKNOWLEDGEMENTS

    Grateful acknowledgements are made to: Dr. Reuschel and
Dr. Keller for stimulating discussion and experimental help;
W. M. Hughes and his group for the FTIR instruments; and
D. F. Greggs, R. R. Lauer, and D. A. Rodamaker for their help
in wafer preparation and analysis.

COMPENSATION EFFECTS IN N.T.D. INDIUM DOPED SILICON

Bernard Pajot and Armand Tardella

Groupe de Physique des Solides de l'E.N.S.
Université Paris VII, Tour 23
75251 Paris Cedex 05, France

ABSTRACT

The infrared absorption spectrum of indium-doped silicon has been investigated in N.T.D. compensated material and compared with the spectrum in the uncompensated material. Besides the decrease in the equilibrium concentration of optically active indium, the presence of ionized donors and acceptors produces a broadening and a shift of the absorption lines, and makes it possible to detect a forbidden transition. Gradual neutralization of the ionized impurities by controlled optical pumping shows the influence of compensation on the detectability of the highly excited levels. The decay with time of the out-of-equilibrium indium absorption is recorded and attributed to pair recombination. Finally, a brief comparison of the optical calibration factors for indium concentration in silicon is given.

INTRODUCTION

Indium has received much attention as a dopant of silicon because its ionization energy of 155 meV makes it useful for extrinsic photoconductivity in the 0.26 eV (4.8 μm) infrared region at comparatively high cryogenic temperatures[1]. This has led to the availability of well characterized In-doped silicon crystals with which significant physical studies could be undertaken. The compensation of this material has been investigated to suppress the thermalization of the residual acceptors with an ionization energy lower than that of indium and neutron transmutation (NT) doping has been used for this purpose. We have demonstrated that NT doping could also be used to calibrate the infrared absorption of indium[2]

independently of electrical measurements; this can be of interest
in measurements of the acceptor concentration in order to determine
the value of physical quantities like the solubility limit. After a
presentation of the indium spectrum in silicon we report here spectroscopic measurements on N.T.D. compensated In-doped silicon as a
function of the intensity of the band-gap pumping, which allow us to
evaluate the influence of the internal Stark effect on the shape of
the line spectrum. These measurements have also revealed the existence of a new line, presumably due to a forbidden transition. The
decay with time of the intensity of the indium lines after switching
off the pumping beam is compared with similar measurements performed
at higher temperature. Finally, we discuss the difference found
between the N.T.D. optical calibration of indium in silicon and the
optical calibration using Hall effect as a primary standard of
indium concentration.

UNCOMPENSATED MATERIAL

    Figure 1 gives an overall view of the $p_{3/2}$ spectrum of indium
in silicon. The difference with previously published spectra[3,4] is

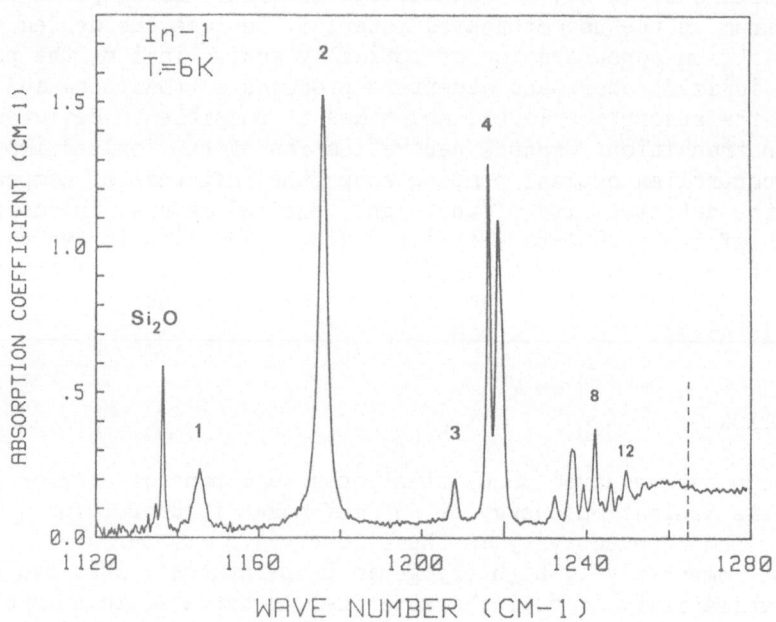

Fig. 1.   Normalized spectrum of indium in silicon ($N_{In} \sim 4 \times 10^{15}$
          at/cm^3). Spectral band pass : 0.35 cm^{-1} at 1250 cm^{-1}. The
          dashed bar at 1265.4 cm^{-1} corresponds to the limit of the
          discrete spectrum.

due mainly to the small indium concentration used ($\sim 4 \times 10^{15}$ atoms/ cm^3) and to the small compensation ratio ($\sim 1 \times 10^{-3}$). The spectral resolution is 0.35 cm^{-1} at 1250 cm^{-1}. The position of the transitions is given in Table 1. Baldereschi and Lipari have calculated[5] the binding energies of the first odd-parity excited levels and it is significant that the doublet 6-6 A corresponds reasonably with the two transitions of which final states are 2 Γ_6^- and 6 Γ_8^-. We have compared the shape of line 2 with a Lorentz profile and this seems to be a reasonable approximation for uncompensated samples, as can be seen from Figure 2. This fit of line 2 was necessary to obtain a reliable value of the integrated intensity of the line for the samples with lower resistivities, as the discrete spectrum is super-imposed to a non-negligible background. This figure also gives an idea of the concentration broadening of the line. This is different from what is observed for line 4 which doubles in width under the same conditions.

Table 1.   Observed position of the indium transitions ($p_{3/2}$) in silicon. Attribution of the final state following Baldereschi and Lipari[5]. Asterisks denote new lines and V.B. the valence band. The previous labeling of lines 10 and 12 is given in parentheses.

Line	Final state	Position cm^{-1}	Line	Position cm^{-1}
1	1 Γ_8^-	1145.5	7	1239.22
2	2 Γ_8^-	1175.9	8	1242.01
3	3 Γ_8^-	1208.07	9 *	1243.80
4	1 Γ_7^-	1216.30	10 (9)	1245.83
4 A	1 Γ_6^-	1218.6	11 *	1247.83
4 B	4 Γ_8^-	1219.5	12 (10)	1249.60
5	5 Γ_8^-	1232.25	13 *	1250.9
6	2 Γ_6^-	1236.25	14 *	1254.1
6 A *	6 Γ_8^-	1237.15	V.B.	1265.4 or 156.9 meV

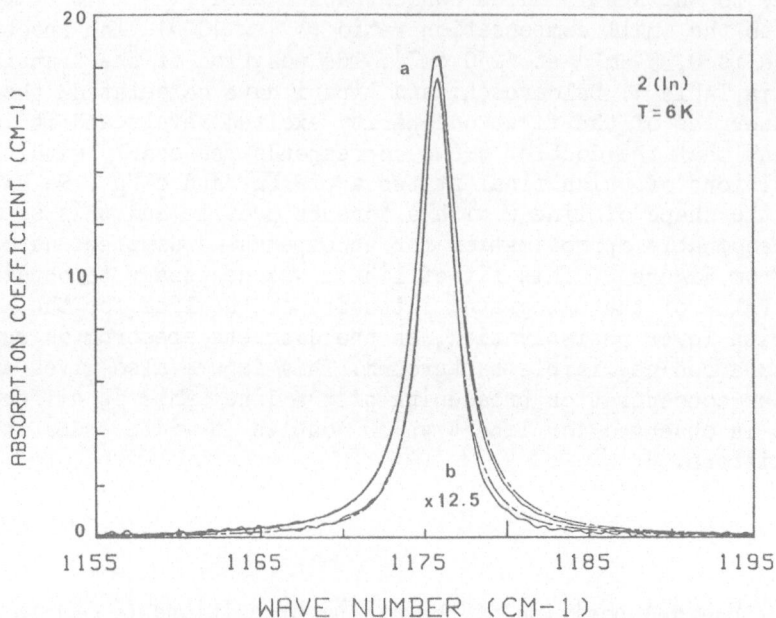

Fig. 2.   Comparison of the experimental shape of line 2 (solid line)
with a Lorentz profile (broken line).   a (outer line) :
$N_{In} \sim 7 \times 10^{16}$ at/cm^3, ordinate scale as indicated.
b (inner line) : [In] $\sim 4 \times 10^{15}$ at/cm^3. Line intensity
increased a factor 12.5 with respect to curve (a).

COMPENSATED MATERIAL

In a semiconductor sample with a net acceptor concentration
$N_A - N_D$ where N_D is the residual unwanted donor concentration, the
deliberate introduction of compensating donors in concentration
$N'_D < N_A - N_D$ reduces the net acceptor concentration to $N_A - (N_D + N'_D)$
If only one kind of acceptor is dominant, the difference between
the integrated intensity of a characteristic spectral line of this
acceptor under thermal equilibrium before and after compensation
must be proportional to the concentration of compensating donors
added if a compensation efficiency of 100 % is assumed. The concur-
rent advantage of NT doping is that the introduction of phosphorous
can be controlled carefully. In the case of In-doped material, the
activation of ^{113}In is a complication arising because of the long
lifetime of metastable ^{114}In. A requisite of this doping technique
is the annealing of the irradiated material to remove the lattice
defects produced during the irradiation. These annealing steps can

have some influence on the isolated substitutional indium concentration if part of the dopant is complexed with another impurity before
or after the compensation process.

At low temperature, the local electric field produced by the
ionized donors and acceptors in compensated material reduces the
lifetime of the excited acceptor states and modifies the details of
the discrete spectrum. The field-induced decrease of the lifetime
of a level depends on its binding energy and the lower the binding
energy, the larger the lifetime reduction[6]. The calculation of the
mean electric field at a neutral impurity site has been determined
in computer simulations of donor and acceptor distributions by a
Monte Carlo technique as the problem is very difficult to solve
exactly.

The In-doped N.T.D. samples investigated here are the same
which had been previously investigated[2]. They had been irradiated
with a thermal neutron fluence of $1.2 \times 10^{19}/cm^2$ (Cd ratio : 4.8;
thermal neutron flux : $4.3 \times 10^{13}/s.cm^2$). They were first annealed
for 10 h at 800° C and re-annealed for 2 h at 900° C. Samples In-T1
([In] = 8.5×10^{15} atoms/cm^3) and In-T2 ([In] = 2.5×10^{15} atoms/cm^3)
in the present report correspond to samples 127-T and A of reference
2, respectively. The observation conditions are different however
(lower spectral band pass, variable temperature facility, more
efficient optical pumping set up and longer cooling time of the
samples). Neutralization of the samples was achieved by side illumination of the sample with the output of a 12 V halogen lamp focussed
on the sample through a narrow band interference filter (80 meV at
half width) centered at 1.24 eV. The power supply of the lamp could
be varied so that partial neutralization of the samples could be
achieved. The correspondence between the labeling of the spectra,
the operating conditions of the lamp and the corresponding neutralization of the samples are given in Table 2. Concerning the intensity

Table 2.   Illumination conditions. They refer to the power
            input (W) of the halogen lamp. J/J (e) is the
            ratio of the intensity of line 2 under a given
            condition to the intensity under complete neutra-
            lization.

		In-T1	In-T2
	Lamp power	J/J (e)	J/J (e)
a	0	0.69	0.42
b	4.5	0.83	0.79
c	8.75	0.90	0.90
d	19.8	0.97	0.99
e	77	1	1

of line 2, we have found a very good agreement with the previous
results. For sample In-T1, however, the intensity under thermal
equilibrium obtained is smaller (7.1 cm^{-2}) than found previously
(7.8 cm^{-2}). Figure 3 shows a survey spectrum of sample In-T1. The
spectrum labeled (e) has been displaced upwards by 0.4 cm^{-1} for
clarity. The background in spectra (b) and (e) is ascribed to the
photoionization spectrum of neutralized phosphorus. Contrary to the
previous results, the indium lines near 1240 cm^{-1} are clearly
observed in (e). The lifetime of the excited states corresponding
to these lines is very sensitive to the presence of residual ionized
impurities. This is better seen in the expanded scale of figure 4.
For spectrum (d), the concentration of ionized indium is only 3 %
of the total indium concentration, but its effect on lines 5, 6-6A,
7 and 8 is clearly discernible. The asymmetric broadening of the
excited levels is accompanied by a shift of their center of gravity
under the influence of the local electric field and the line is no
longer Lorentzian. This effect is not important and its observation
necessitates a good reproducibility of the results. For line 2, the

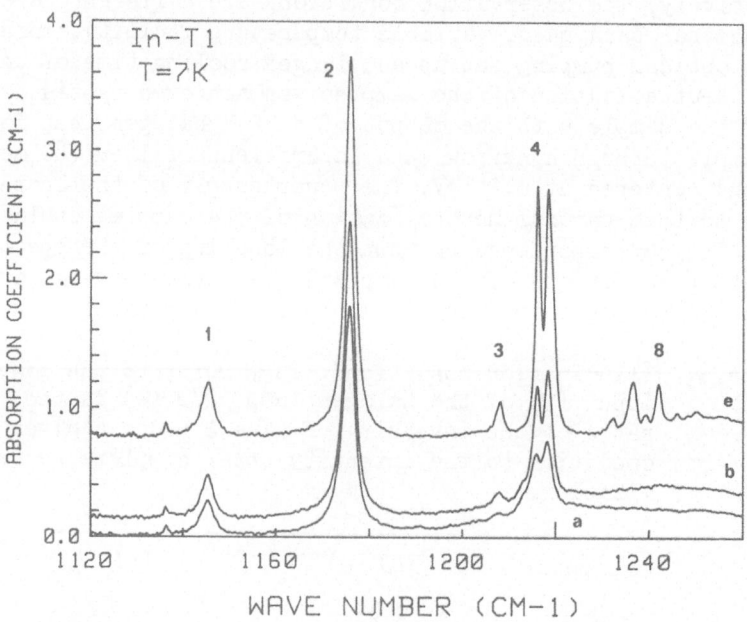

Fig. 3.   Spectrum of N.T.D. compensated In-doped silicon under
          different illumination conditions. $N_p = 2 \times 10^{15}$ atoms/cm^3,
          $N_{In} \sim 8 \times 10^{15}$ atoms/cm^3. The labeling of the spectra
          refers to the illumination conditions given in Table 2.
          Spectrum e is shifted upwards by 0.4 cm^{-1}.

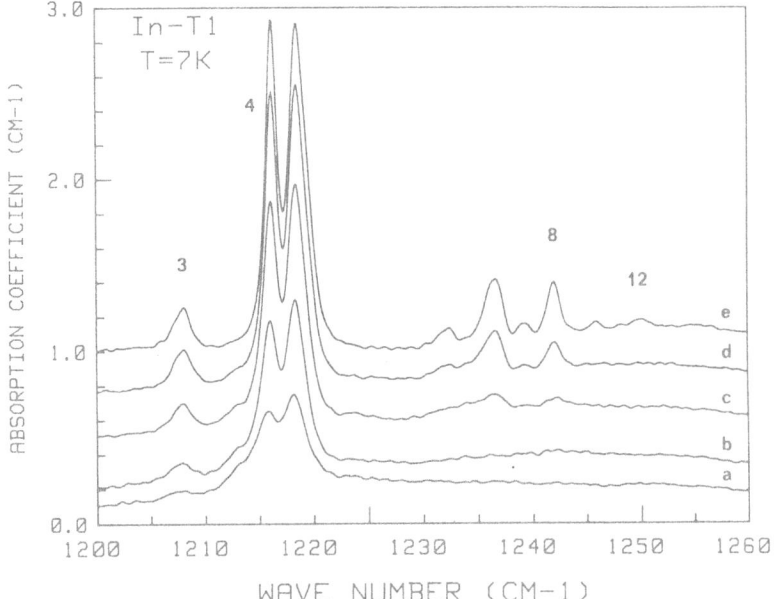

Fig. 4.   Spectrum of N.T.D. compensated sample of Fig. 3 under
          different illumination conditions. The shoulder at
          1213 cm^{-1} is ascribed to a field-induced forbidden
          transition. Spectra c, d and e are shifted upwards by
          0.2, 0.4 and 0.6 cm^{-1}, respectively.

total shift between (a) and (e) is (0.24 ± 0.08) cm^{-1} or 0.03 meV
for both samples. For line 4, the shift amounts to 0.4 cm^{-1} for
sample In-T1.

    Besides the broadening and shift of the lines, we have observed
in the two compensated samples a new line in the spectra obtained
when ionized impurities are present. This line is observed as a
shoulder of line 4 at $\sim$ 1213 cm^{-1} in Figure 4, but it was observed
as a well resolved line in sample In-T2. This line can be observed
as a shoulder in the indium spectrum shown in the paper by Onton et
al.[3], indicating that the sample used was compensated. Lipari et al.
have determined experimentally from the two-hole luminescence
spectrum and theoretically the binding energy of the first even-
parity states of B, Ga and In in silicon[7]. The energy of the forbidden 1 $\Gamma_8^+ \to$ 3 Γ_8^+ transition of indium is 1200 cm^{-1} (149 meV) and
this is near the new line observed at 1213 cm^{-1}. The absence of
evidence for another n Γ_8^+ line suggests however that this line
could be a 1 $\Gamma_8^+ \to$ 2 Γ_7^+ transition.

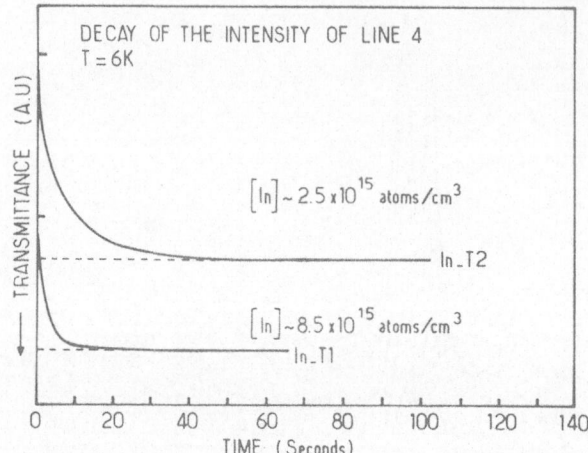

Fig. 5. Recorder trace (transmittance) of the decay with time
         of the intensity of line 4. The trace for In-T1 is
         normally <u>above</u> that for IN-T2, but it has been shifted
         for clarity.

We have found that ionized acceptor impurities are totally
neutralized by band gap light pumping in p- and n-type material.
When the excitation beam is switched off, the system comes back to
the initial equilibrium. We have recorded the time decay of the
intensity of line 4 with time at T = 6 K for samples In-T1 and In-T2
(Figure 5). There is a net correlation between the indium concentration and the decay rate. Sundström et al. have assumed that the
post-excitation decay of the continuum absorption of indium they
observe in P-compensated In-doped silicon comes from the capture
of free electrons by neutral indium[8]. As a proof, they produce a
luminescence spectrum they attribute to the above radiative capture.
The spectral dependence of this luminescence is actually due to
radiative In-P recombination[9] and we believe that this is also the
origin of the decay of line 4 presented in Figure 5.

We note that the integrated intensity of line 2 in sample
In-T2 under equilibrium <u>before</u> transmutation was lower than <u>after</u>
transmutation under band gap light pumping. We have re-checked the
compensation of sample In-T2 before NT doping using an adjacent
sample and found it undetectable on the intensity of line 2. So,
the only explanation we can offer is that during the compensation
steps, a pre-existing electrically inactive indium complex is

dissolved, increasing the substitutional indium concentration. For sample In-T1, there was practically no variation of the <u>total</u> In concentration before and after NT doping, but the difference between the integrated intensities is 3.2 cm^{-2} against 2.5 cm^{-2} for sample In-T2. These figures are averaged on several independent measurements and they are considered reliable to better than ± 0.2 cm^{-2}.

For calibration purpose, we believe that the factor 8 x 10^{14} In atoms/cm^3/cm^{-2} given in reference 2, obtained from a decrease of 2.5 cm^{-2} in the integrated intensity of line 2 after the introduction of 2 x 10^{15} P atoms/cm^3 by neutron transmutation doping is more realistic than the value of 2 x 10^{15} In atoms/cm^3/cm^{-2} given by Jones et al. as deduced from electrical measurements[10]. There are two possible reasons for this discrepancy : i) When relating hole concentration at room temperature to an indium concentration, one must use the ionization energy of indium <u>at room temperature</u>, which is lower than the low temperature value. We estimate the room temperature valve to be 140 meV[11]. ii) When the hole concentration is deduced from room temperature Hall measurements at low field, the Hall factor must be taken into account and a value of 0.8 for this factor seems to be realistic[12], at least for samples with p $\gtrsim$ 1 x 10^{15} cm^{-3}. With such corrections, the calibration factor in ref. 10 would be lowered to a figure not too different from our N.T.D. optical factor.

ACKNOWLEDGEMENTS

The authors wish to thank D. Roche from C.E.N.G. for the NT doping of the present samples. The assistance of S. Squelard for the annealing is also gratefully acknowledged.

REFERENCES

1. T. T. Braggins, H.M. Hobgood, J.C. Swartz, and R.N. Thomas, High infrared responsivity Indium-doped silicon detector material compensated by neutron transmutation, I.E.E.E. Transactions Electr. Dev. ED-27:2 (1980).
2. B. Pajot, D. Débarre, and D. Roche, Neutron transmutation as a method to calibrate the infrared absorption of indium in silicon, <u>J. Appl. Phys.</u> 52:5774 (1981).
3. A. Onton, P. Fisher, and A.K. Ramdas, Spectroscopic investigations of group-III acceptor states in silicon, <u>Phys. Rev.</u> 163:686 (1967).
4. B. C. Covington, R.J. Harris, and R.J. Spry, Observation of additional excited-state lines of indium in silicon, <u>Phys. Rev. B</u> 22:778 (1980).

5. A. Baldereschi and N.O. Lipari, Interpretation of acceptor
    spectra in Si and Ge, in: "Proc. 13th Conf. Phys. Semi-
    conductors", F.G. Fumi, Ed. Tipografia Marves, Rome (1976).
6. P. R. Bratt, Impurity germanium and silicon infrared detectors,
    in: "Semiconductors and Semimetals, Vol. 12", R. K.
    Willardson and A. C. Beer, Eds. Academic Press, New York
    (1977).
7. N. Lipari, M.L.W. Thewalt, W. Andreoni, and A. Baldereschi,
    Central cell effects in the shallow acceptor spectra of Si
    and Ge, J. Phys. Soc. Japan 49:suppl. A 165 (1980).
8. B. O. Sunström, L. Huldt, and N.G. Nilsson, Photoinduced infra-
    red absorption and luminescence in Indium-doped silicon,
    Physica Scripta 18:414 (1978).
9. U. O. Ziemelis and R.R. Parsons, Sharp line donor-acceptor pair
    luminescence in silicon, Can. J. Phys. 59:784 (1981).
10. C. E. Jones, D. Schafer, W. Scott, and R.J. Hager, Carbon-
    acceptor pair centers (X centers) in silicon, J. Appl. Phys.
    52:5148 (1981).
11. A. Tardella and B. Pajot, The infrared spectrum of indium in silicon
    revisited, J. Physique 43:1789 (1982)
12. W. R. Thurber, R.L. Mattis, Y.M. Liu, and J.J. Filliben,
    Resistivity-dopant density relationship for boron-doped
    silicon, J. Electrochem. Soc. 127:2291 (1980).

CORRELATION OF NTD-SILICON ROD AND SLICE RESISTIVITY

W. Michael Wolverton

Electronics Division
Monsanto Electronic Materials Company
St. Peters, MO 63376

INTRODUCTION

One of the most important parameters of electronic materials is resistivity. Silicon suppliers are interested in controlling dopant incorporation and characterizing the electronic behavior of ingots and slices. Customers of electronic-grade silicon are interested in understanding how resistivity level and uniformity affect device yields. Resistivity and sheet resistance are used as standards for quality control by both suppliers and customers.

Neutron transmutation doped (NTD) silicon is an electronic material which presents an opportunity to explore a high level of resistivity characterization. This is due to its excellent uniformity of dopant concentration. Appropriate resistivity measurements on the ingot raw material can be used as a predictor of slice resistivity.

Each arrow in the following sequence shows a resistivity correlation point for a process step in the manufacture of NTD-silicon slices.

ZONE REFINED ROD → NTD ROD → AS-CUT SLICE → AS-SOLD SLICE

Correlation of finished NTD rod (i.e. ingot) resistivity to as-cut slice resistivity (after the sawing process) is addressed in the scope of this paper.

Part of the slice thickness is electrically inactive. The inactive portion is presumably due to the structural damage and stresses inflicted near the slice surface by the sawing process.

An inactive thickness of 10 μm has been found experimentally[1].
The correlation of rod to slice resistivity can be influenced by
this electrically inactive layer.

The problem is an erratic shift of slice resistivity compared
to rod resistivity. The latter is typically measured by positioning a collinear four-point probe (FFP) at the centers of the rod
end faces. Figure 1 shows how slices from four of eleven test rods
shift significantly upward in resistivity.

Good correlating equations are needed to determine those rod
resistivity specifications which are the best predictors of the
slice resistivities ordered by customers. The objective is not
just to minimize the number of slices which are rejected for too
high or too low a resistivity. The main objective is to predict the
amount of rod doping (from neutron irradiation) which is needed to
produce <u>slices</u> having resistivities distributed around the middle
of very tight resistivity specifications.

THEORY

Resistivity is a measurement (of a material property) which is
used in theoretical descriptions of how electrons flow through material. It is affected not only by chemical impurities, but also
by physical defects and inhomogeneous dopant incorporation. The
basic resistivity equations are premised on the assumption that the
piece of silicon has perfectly uniform resistivity.[2,3,4] Other
important assumptions mentioned by Valdes[4], are:

1. Any injected minority carriers recombine near the elec-
    trodes. (Lapped surfaces facilitate this recombination
    so that the resistivity measurement is not affected).
2. The boundary (i.e. the surface of the material) is either
    nonconducting or perfectly conducting.
3. Probe spacing should be much greater than the diameter of
    the metallic probe tip.

Figure 2 summarizes the resistivity techniques which are commonly used in rod and slice measurements. The FPP slice-resistivity equation shows that resistivity is directly proportional to
thickness. In practice this means that twice as much direct current is passed through the outer two probes for a 0.4 mm thick slice
as for a 0.2 mm thick slice. If these two slices were cut from a
rod having uniform resistivity, the voltage measured between the
two inner probes would theoretically be the same for both slices,
indicating identical resistivity. However, let us assume that both
slices have a measured thickness which is 0.01 mils greater than the
electrically active thickness. The resistivity would then have a
value which measured $(0.40 \div 0.39 - 1) \times 100 = 2.6\%$ too high for the
0.4 mm thick slice and 5.3% too high for the 0.2 mm thick slice.

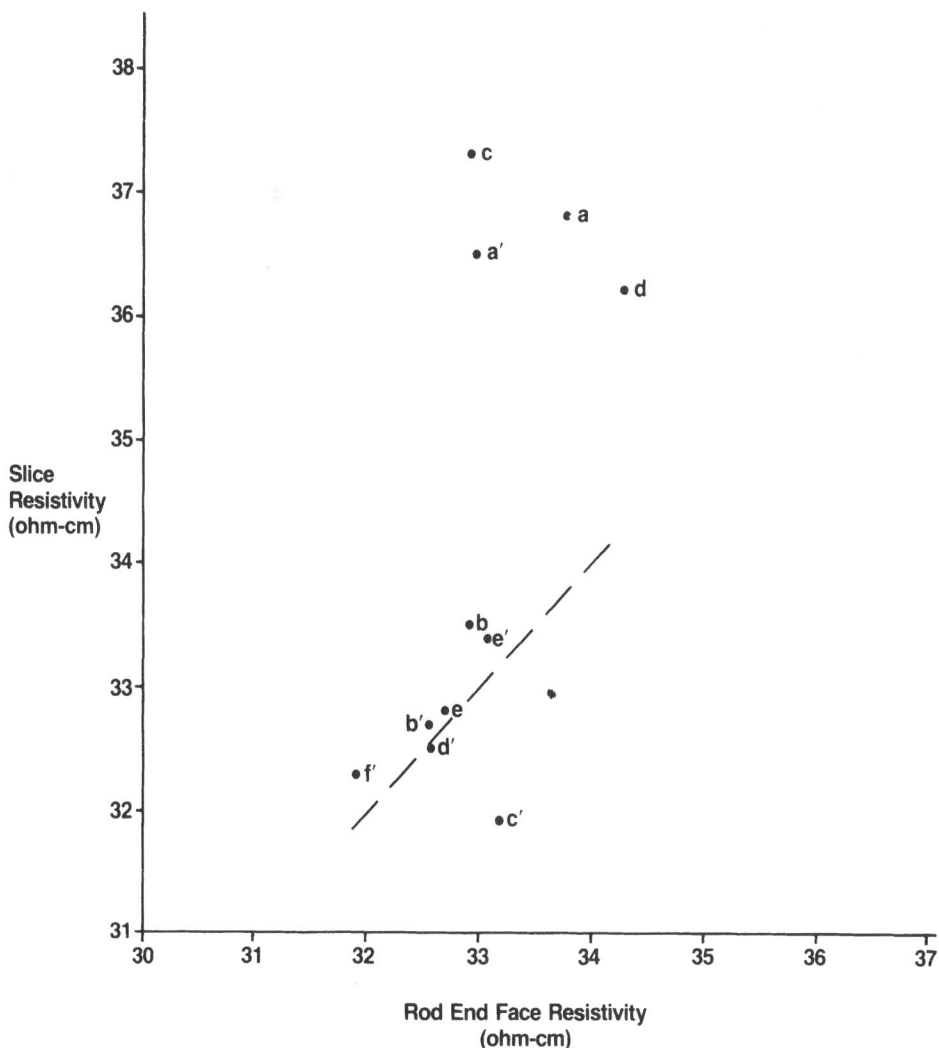

Fig. 1.   Average four-point probe (FPP) resistivity at slice half
           radius (midway between the center and the edge of the slice
           versus average FPP resistivity measured at the centers of
           the two rod end faces, for eleven rods.

4-Point Probe:

$$\rho_{slice} = \frac{V}{I} \frac{\pi \, t}{ln2}$$

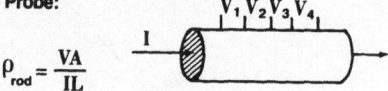

$$\rho_{rod} = \frac{V}{I} \, 2\pi s$$

ρ = Resistivity (ohm-cm)
V = Voltage (Volts)
I = Current (Amps)
t = Thickness (cm)
s = Probe Spacing Of Tips (cm)

2-Point Probe:

$$\rho_{rod} = \frac{VA}{IL}$$

A = Cross-Sectional Area Of Rod (cm²)
L = Length Of Rod (cm)

Fig. 2.   Resistivity probe positions for rod and slice measurements.
Vertical dots show probe-tip positions on the center of
a rod end face. This is termed "rod end center".
Horizontal dots show probe tip positions for FFP "rod
profile" readings. V_1, V_2, V_3 and V_4 are two-point probe
rod profile readings at the points indicated.

It is apparent that inactive thickness becomes more critical to accurate resistivity measurements as slice thickness is reduced.

Correction factors become increasingly necessary as the FPP
measurement is taken closer to the edge of the slice. These correction factors depend on the probe spacing. They are derived by assuming boundaries are nonconducting[5].

In addition to the four-point probe (FPP), the two-point probe
for rods (see Figure2) and the five-point probe for slices have the
potential to improve correlation between rod and slice resistivity.
These probe techniques are described elsewhere[6,7]. The five-point
probe technique adds a DC bias to the FPP in order to reduce the contact resistance on high-resistivity, high-lifetime silicon. This
turns the current injection problem from rectifying contacts into an
advantage by controlling the slope of the current-voltage curve.

Resistivity radial gradient (RRG) for rods is commonly defined
in terms of FPP measurements taken at an end-face position which is

Table 1. NTD Silicon Rod and Slice Resistivity:
Some Data for .33mm Thick Slices.

	QTY.	$\bar{x}$	s (pooled)
RRG of Rods (%)	11	−.05	1.53
RRG of Slices (%)	550	.54	1.05
Slice ρ_c 4-Point (ohm-cm)	300	33.03	.65
Slice ρ_c Noncontact (ohm-cm)	300	33.27	.78

KEY: c = center
r = radius
RRG = $(\rho_r - \rho_c) \div \rho_c$
      $\frac{}{2}$

not at the rod-end center. For example, the RRG for a half-radius
position is shown in the key to Table 1.
A new resistivity radial gradient (RRG) is defined for rods as follows.

RRG´ ≡ Average Profile/Average End Center (1)

Empirical data are needed to test the following hypothesis.
The shift of slice-center resistivity compared to rod-end center resistivity is a function of RRG´.

Shift = f(RRG´)   (2)

In other words, is the shift illustrated in Fig. 1 a function
of RGG´? Essentially what is being tested here is whether or not
inhomogeneous dopant incorporation can be shown to affect FPP slice
resistivity readings. The contention is also that such inhomogenei-
ties can include physical defects and not just the well known chemi-
can impurity dopants.

This contention is supported by references in the literature.
High energy radiation can produce lattice defects whose energy
levels can result in either acceptor or donor behaviour. Divacancies
can act like acceptor defects[9]. Dislocations can act like acceptors;
about twenty million acceptors can be introduced per dislocation[10].

The above defects can be viewed as point defects (zero-
dimensional) or line defects (one-dimensional). Kazmerski[11] points
out that current conduction is blocked in going across grain boundaries (two dimensional). However, it has been shown for a germanium
bicrystal that conductivity along a grain boundary is high. This is
due to banding of defect states in the boundary. Perhaps there is
similar conduction along dislocations.

Pressure reduces the resistivity of N-type silicon[12]. One
might infer that resistivity is therefore affected by the lattice
strain which is associated with internal stresses in the material.

The mechanisms by which the various physical defects are considered to affect resistivity can be seen in equation (3)[8]. The
resistivity, ρ, of an N-type semiconductor is

$$\rho = \frac{1}{q\mu_n N}$$ (3)

where q is the electronic charge, μ_n is the electron mobility and N
is the concentration of electrons. The various physical defects
would change the resistivity by affecting carrier concentration
and/or mobility. Mobility is a function of temperature and may
increase along dislocations.

EXPERIMENTAL WORK

The objectives of the present experiment were to test the hypothesis represented by equation (2) and to achieve the manufacturing economies, efficiencies and process capabilities which were mentioned in the introduction.

The resistivities of eleven (11) rods and fifty (50) slices per rod were measured just before and just after the sawing plus cleaning process. The eleven float-zone refined rods were doped by means of neutron irradiation in the ten megawatt, light water, research reactor at the University of Missouri in Columbia, MO. The rods were 2.00 inches in diameter.

The most accurate resistivity measurement methods were used, in order to provide sufficient precision to distinguish small differences in carrier concentration. The following procedures of high precision resistivity measurements were practiced.

1) The appropriate resistivity equations were used with temperature and other correction factors taken into account. Actual measured slice thicknesses ($\pm$ 5 μm) were input into the resistivity calculation.

2) Probe tip pressure, spacing, alignment, diameter and visual condition were monitored and controlled.

3) Rod surfaces were sandblasted prior to the resistivity measurements.

4) The direction of the current was reversed and the measurements repeated in order to check for P-N junctions and Schottky barriers.

5) Current input which provide zero slope on the resistivity versus current curves[13] were used.

6) Calibrations against resistivity standards were performed daily.

Two techniques which influence accuracy were not used, but they deserve to be mentioned. A two-configuration technique reportedly[14] reduces error from nonequal probe spacing and reduces the need for correction factors near the edge of a slice. A probe whose four tips are positioned in a square[15] is considered to be more sensitive to resistivity anisotropies than the conventional collinear positions of the tips.

The fifty slices per rod were randomly selected by thickness sorting all of the slices for a given rod. This was done prior to measuring resistivity. The nominal slice thickness was 0.33 mm.

Average resistivities were used in the computations leading
to the graphs of Figures 1, 3, 4, 5 and 6. For example on the
y-axis of Figure 3, the average of FPP measurements taken at the
centers of fifty slices is compared to the average of two FPP end-
center measurements taken on either end of the rod from which the
slices were cut. On the x-axis the average of the rod-profile readings is compared to the average of the end-center readings.

Two-point and four-point profile readings were taken one centi-
meter apart along the axis of the rod. Average profile readings do
not include any readings closer than two cm to the end of a rod.

DISCUSSION

Table 1 shows that, for 300 slices, the average slice resistivity measured by FPP agrees with the average of resistivity readings
measured with a noncontact gauge to within one percent.

Figure 3 includes data from the first five rods tested. In this
first set of rods, the rods are labelled a,b,c,d,e. It is apparent
that excellent correlation exists between the shift (of slice center
resistivity compared to rod-end center reisstivity) and RRG´. The
best correlation is obtained by using an average of the two-point
and the four-point rod profile.

Note that there is a 3.1% shift where RRG´ equals zero. Coin-
cidentally, 3.1% of the 0.33 mm nominal slice thickness equals 10 μm
which is a typical electrically-inactive thickness. Therefore,
slices cut from a rod having a perfectly uniform resistivity (i.e.
RRG´=0) would apparently shift upward in resistivity due to inactive
thickness.

In other words, slices of the same nominal thickness can be
used to determine inactive thickness by means of rod and slice resistivity data. This approach avoids the waste which would occur
if slices having a wide range of thicknesses were required to obtain inactive thickness.

Let SHIFT% be the shift where RRG´ equals zero. Also, define
RRG´ critical% to be the RRG´ where the shift equals zero. In Figure 3 it is apparent that:

$$- \text{SHIFT \%} = \text{RRG}´ \text{ critical \%} \qquad (4)$$

Figure 4 includes data from the next six rods tested. In this second set of rods, the rods are labelled a´,b´,c´,d´,e´,f´. Within the domain of the data of Figure 3 there is good agreement with the first set of five rods. However the linear relationship between shift and RRG´ seems to break down for RRG´ less than RRG´ critical. Let us use the definition that all rods with RRG´ greater than RRG´ critical are in domain "A". All rods with RRG´ less than RRG´ critical are in domain "B".

In Figures 5 and 6, a line is drawn around the rods in domain "B". The other rods are in domain "A".

Note that rods a, c, d and a´ are in domain "A". Figure 1 shows that these rods have slice resistivities which are significantly higher than might be expected from a knowledge of only the rod end-center resistivities. In contrast, Figure 5 and 6 show that slices cut from domain "A" have resistivities which correlate well with rod-profile resistivities.

The slices cut from rods in domain "B" have poor correlation with rod profile resistivities. For these rods the conventional practice of using rod end-center readings to predict slice resistivity is preferred (see dashed line in Figure 1).

In practice either two-point or four-point rod profile resistivities can be specified by means of the correlating equations in Figures 5 and 6. These equations can be used to improve the prediction of slice resistivity for rods in domain "A". Manufacturing economies have resulted from the new, empirically-derived rod specifications. Neutron doping targets have been adjusted to improve the capability of producing slices with resistivities closer to the middle of slice resistivity specifications.

If as-cut slices could be produced with zero electrically inactive thickness then according to equation (4), the critical radial gradient would also equal zero. If data were obtained for such slices and presented as shown in Figure 3, the resulting correlation line would be expected to shift downward so that SHIFT% would equal zero% instead of 3.1%. Modified slices could be tested to verify the validity of equation (4) in this manner.

The value of equation (4) is in the potential to determine optimal rod specifications and neutron doping targets for different sawing processes and for slices of different nominal thickness.

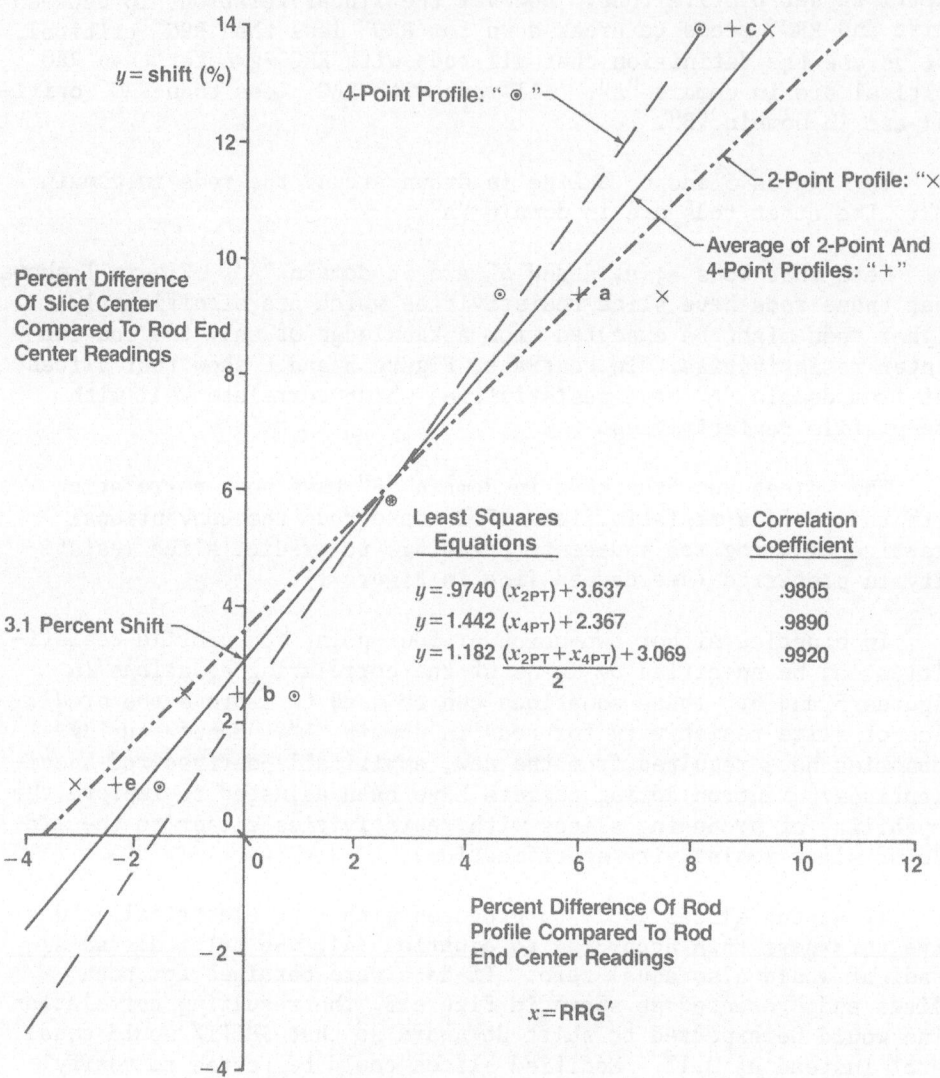

$$y = .9740\,(x_{2PT}) + 3.637 \qquad .9805$$
$$y = 1.442\,(x_{4PT}) + 2.367 \qquad .9890$$
$$y = 1.182\,\frac{(x_{2PT} + x_{4PT})}{2} + 3.069 \qquad .9920$$

Fig. 3.   Resistivity Shift versus Resistivity Radial Gradient
           (RRG´). Refer to equations (1) and (2). Data from
           the five rods in the first set of rods tested are
           plotted.

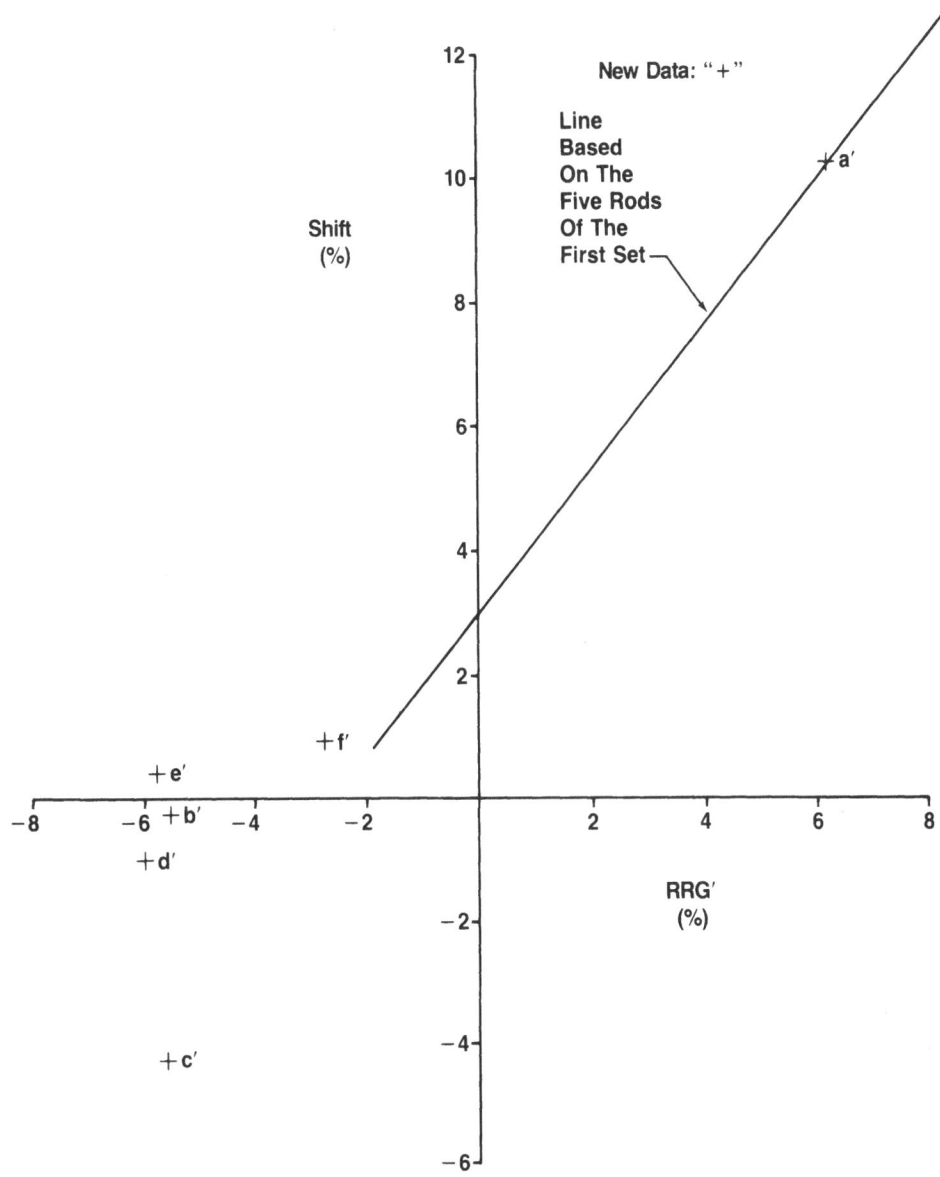

Fig. 4.   Resistivity Shift versus Resistivity Radial Gradient
          (RRG´). Refer to equations (1) and (2). Data from
          the six rods in the second set of rods tested are
          shown by "+".

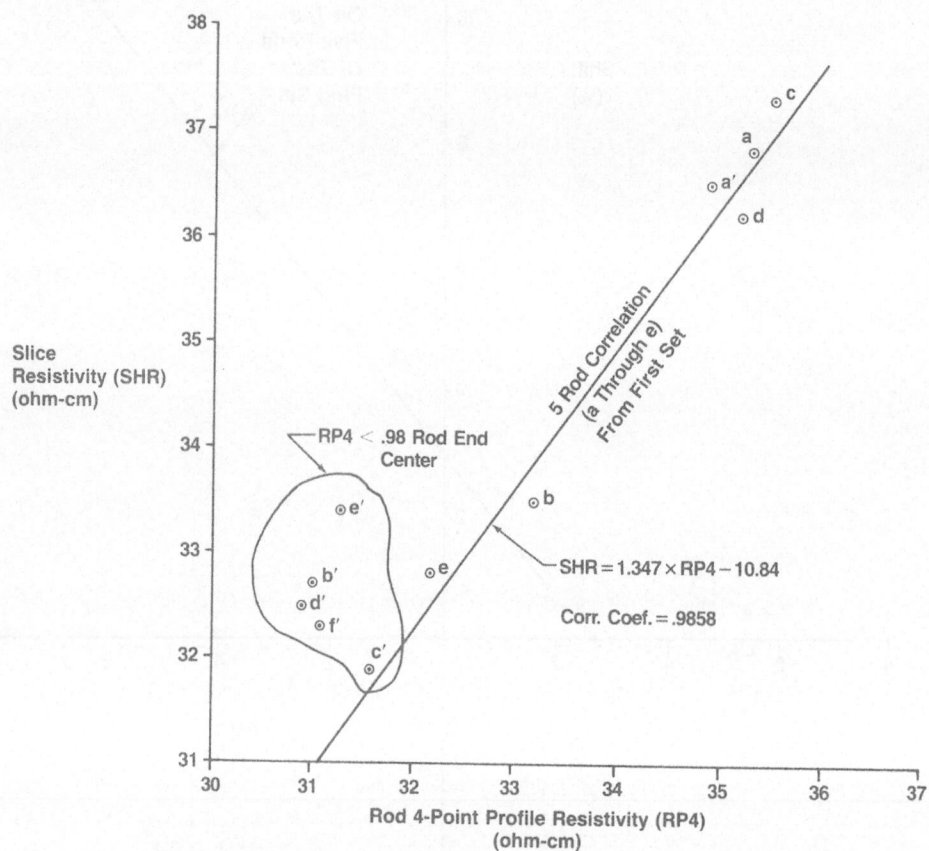

Fig. 5.   Average slice half radius resistivity (SHR) versus
           average rod FPP resistivity (RP4) for all eleven rods.

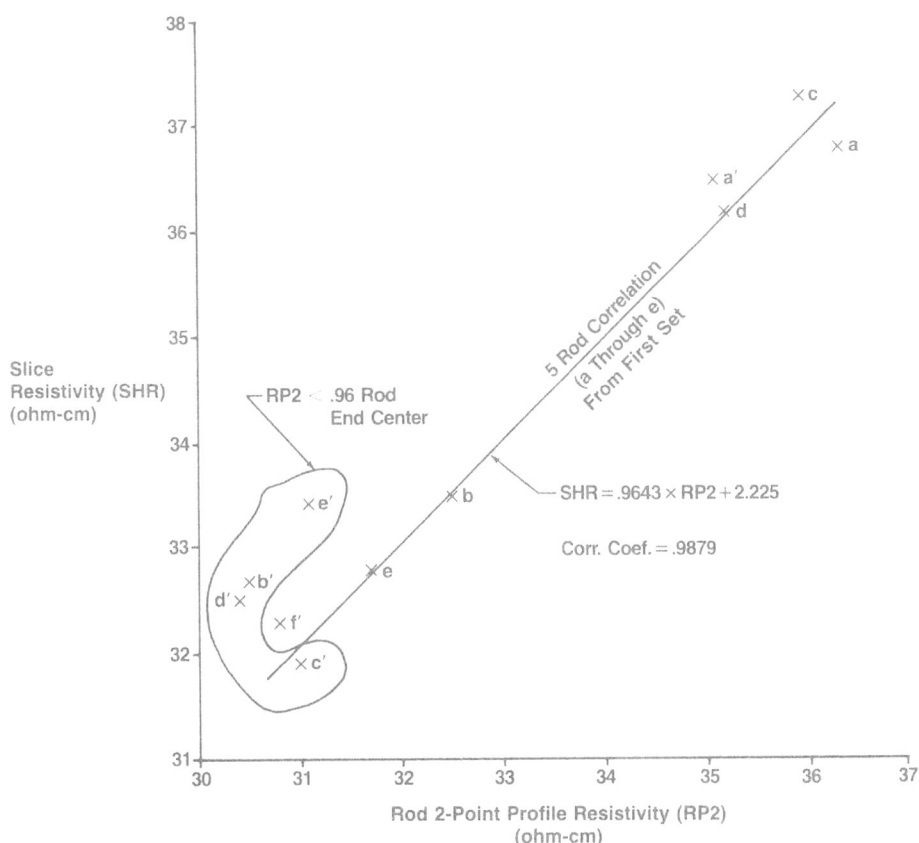

Fig. 6.   Average slice half radius resistivity (SHR) versus
average rod two-point profile resistivity (RP2)
for all eleven rods.

CONCLUSIONS

    1) Empirical data show that the shift of slice-center resisti-
vity compared to rod-end center resistivity is a function of a new
kind of rod radial-resistivity gradient, RRG´. The function has
two domains. Measurement of FPP resistivity at various positions on
the rod surface determine if the rod is in domain "A" or domain "B".
Most rods are in domain "A".

    2) Correlating equations show how to significantly improve the
prediction of slice resistivity for rods in domain "A". Optimal
rod-resistivity specifications and fine tuning of neutron doping
targets can be obtained from the equations.

    3) An unexpected finding is that the electrically inactive
thickness (which is related to the sawing process) can apparently
be derived from resistivity measurements on rods and slices.

    4) The theoretical explanation for the empirically inferred
relationships in equations (2) and (4) is not complete. Zero-,
one- and two-dimensional physical defects are expected to influence
resistivity based on experimental results reported by others. It is
suggested that this influence could be generally understood in terms
of lattice strain and internal stresses in the material.

    5) The new rod resistivity specifications have resulted in
manufacturing economies in the production of NTD-silicon slices.
The quality control system for data collection has resulted in an
improved capability to achieve resistivities closer to the middle
of the tight, NTD-silicon slice-resistivity specifications.

ACKNOWLEDGEMENTS

    Special thanks are extended to Harry Woolsey and those in the
St. Peters Quality Control Laboratories who helped with the
experimental work.

    The author would like to acknowledge helpful discussions with
Horst Kramer and Scott Jordan.

REFERENCES

1. O. Malmros, Precision Resistivity Measurements On NTD-Silicon,
   in: "Neutron-Transmutation-Doped Silicon," J. Guldberg, ed,
   Plenum Press, New York (1981) p. 411.
2. M.P. Albert and J.F. Combs, Correction Factors for Radial
   Resistivity Gradient Evaluation of Semiconductor Slices,
   IEEE Trans. Electron Devices, (April, 1964), 148.
3. "Geometric Factors In Four-Point Resistivity Measurements,"
   Technical Report TR18A, Geoscience Instruments Corp. (1966).

4. Valdes, Resistivity Measurements on Germanium for Transistors, Proc. I.R.E. 42:420 (1954).

5. F.M. Smits, Bell System Tech J. 37:711 (1958).

6. P.F. Kane and G.B. Larrabee, "Characterization of Semiconductor Materials," McGraw Hill, New York (1970)

7. E.G. Bylander, "Materials for Semiconductor Functions," Hayden Book Co., New York (1971) pp. 53-54.

8. A.S. Grove, "Physics and Technology of Semiconductor Devices," John Wiley and Sons, New York (1967) pp. 112,143.

9. N.N. Gerasimenko and K.B. Tnyshtykbaev, Formation of Radiation Defects In Silicon Containing Hydrogen Atoms, Sov. Phys. Semi. 14:9 (1980).

10. H. Ono and K. Sumino, Acceptors In N-Type Silicon Crystals Induced By Plastic Deformation, Jap. J. Appl. Phys. 19:10 (1980).

11. L.L. Kazmerski, "Polycrystalline and Amorphous Thin Films and Devices," Academic Press, New York (1980) pp. 163-164.

12. W.R. Runyan, "Silicon Semiconductor Technology," McGraw Hill, New York (1965) p. 181.

13. L.H. Garrison, Proper Current Input for True Resistivity Measurements of Single Crystal Silicon Both N- and P- Type and Single Crystal Slices, Sol. State Tech. (May 1966) p. 47.

14. D.S. Perloff et al., Dose Accuracy and Doping Uniformity of Ion Implantation Equipment, Sol. State Tech. (Feb., 1981) p. 112.

15. J.D. Wasscher, Note On Four-Point Resistivity Measurements On Anisotropic Conductors, Philips Res. Repts. 16:4 (1961).

A DETAILED ANNEALING STUDY OF NTD SILICON

UTILIZING RAMAN SCATTERING

H.R. Chandrasekhar[*], Meera Chandrasekhar, J.M. Meese
and S.L. Thaler[**]

Department of Physics and Research Reactor Facility
University of Missouri-Columbia, Columbia, MO 65211, USA

ABSTRACT

The technique of Neutron Transmutation Doping (NTD) of a silicon crystal by an in-situ transmutation of the ^{30}Si isotope into ^{31}P via the reaction

$$^{30}Si(n,\gamma) \ ^{31}Si \rightarrow \ ^{31}P + \beta^-$$

has distinct advantages over conventional doping schemes. Among these are the precise ability to control doping concentrations and the overall uniformity of doping. The fast-neutron component of the reactor spectrum produces damage in the crystal which can be repaired by an annealing process.

A detailed isochronal annealing study of the vibrational spectrum of NTD silicon has been carried out by means of Raman scattering. The peaks introduced into the vibrational spectrum at low (120 cm^{-1}), intermediate (320 cm^{-1}) and high (490 cm^{-1}) frequencies, due to radiation damage, have been monitored in the annealing range from 30° to 900°C.

The isochronal annealing data of this study was adjusted via a kinetic model, to allow for a comparison with annealing curves for specific point defects, obtained utilizing electron spin resonance. Because of the close correlation between the ESR and

[*]Alfred P. Sloan Foundation Fellow.
[**]Now at McDonnell Douglas Corp., St. Louis, Mo.

Raman annealing data, tentative peak assignments have been made. It is concluded that the defect mode appearing at 120 cm^{-1}, immediately following room temperature irradiation, is associated with the 4~vacancy (P~3) and that the mode appearing at 128 cm^{-1} after the 350°C anneal, is due to the 5~vacancy (P~1).

It was observed that the Raman intensity appears to follow separate second~order decays in both the low (30°~300°C) and high (350°~600°C) temperature ranges, possibly indicating the uncorrelated nature of the combining defects. The low activation energies obtained (roughly 0.5 and 1.0 eV respectively), as well as the low attempt frequencies calculated (10^6 sec^{-1}), suggests a rate~limiting step involving the ionization of a carrier from the defects. The fact that the free carrier recovery of the extrinsic sample nearly parallels the defect annealing, tends to corroborate this point of view.

Further, it is noted that the resistivity data may be fit to a single, second~order decay over temperature range from 30° to 600°C, from which an activation barrier of 0.65 eV may be calculated. This fact may indicate that either similar or identical defects dominate the annealing kinetics over this entire temperature range.

1. INTRODUCTION

Recently, the vibrational spectrum of NTD silicon was observed by Chandrasekhar et al. (1981). In that study, it was found that basically three new defect~induced modes appear in the Raman spectrum, at low (120 cm^{-1}), at intermediate (320 cm^{-1}), and at high (490 cm^{-1}) frequency (Fig. 1). All three damage features are observed in the Γ_1 scattering geometry of crystalline silicon and have been demonstrated to be first~order features. Further, all three peaks, observed for both irradiated float~zone (250 Ω~cm) and boron~doped (0.001 Ω~cm) silicon, appear to scale with neutron dose.

Preliminary data on the annealing behavior of the neutron damage has shown that all three of the above damage peaks decrease simultaneously in the annealing stages between 200° and 300°C, with the low frequency peak shifting upward in frequency. In contrast, the intermediate and high frequency peaks remain relatively static in position.

In order to account for the contrasting behavior of the low and higher frequency damage peaks, the heuristic model of Brodsky (1975), has been applied. In this model, the introduction of a defect effectively increases the dimension of the primitive cell. The Brillouin Zone (BZ) is thus folded into the zone

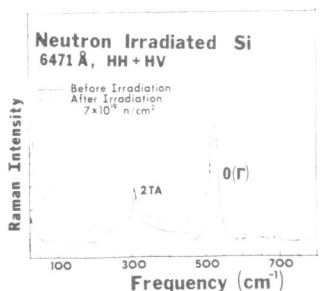

Fig. 1. Raman spectra of silicon, before and after irradiation

Center, giving rise to new first-order processes, represented by the three damage peaks reported above. Since the optic branches are fairly flat, the zone folding does not significantly alter the position of the optic-derived, high frequency peak. On the other hand, unfolding the BZ (corresponding to annealing the damaged crystal) qualitatively shows the shift of the low frequency peak toward higher frequency, due to the sloping nature of the acoustic branches.

In the annealing study which has been performed, much attention has been given to the dynamic nature of the low-frequency peak, through a computer enhancement of the digitized spectra. These Raman spectra were obtained at $300^{\circ}K$, using 300 mW of 6471 Å radiation from a Kr^+ laser, a double monochromator with holographic gratings, and a cooled photomultiplier with photon counting electronics. Isochronal anneals of the p-type sample (0.001 Ω-cm), irradiated to a fluence of 7×10^{19} neutrons/cm^2, were carried out for 30 minutes, under flowing nitrogen gas. The results of this study are discussed below.

2. ANALYSIS OF THE ANNEALING SPECTRA

The observed spectra for all of the annealing stages were manipulated in the following way. First, the background was subtracted from the Γ_1 damage spectrum and the result multiplied by the factor $\omega / \{\omega_L - \omega)^4 [n(\omega, T) + 1]\}$ (where ω_L is the incident laser frequency and $n(\omega, T)$, the Bose-Einstein population factor),

so as to yield the Raman effective density of states, as
prescribed by Shuker and Gammon (1970).   Second, a signal
averaging was performed, prior to vertical enlargement of the low-
frequency region from 0 to 300 cm^{-1}.   The results are shown in
Figure 2.

It was also possible with the computer enhancement, to follow
peak intensity through the annealing sequence.   In Fig. 3,
normalized intensity data is shown for both the low and high
frequency peaks, not only for the computer-manipulated data for
the p-type silicon (p-Si) sample, but also for the n-type silicon
(n-Si) sample studied previously by Chandrasekhar et al. It is to

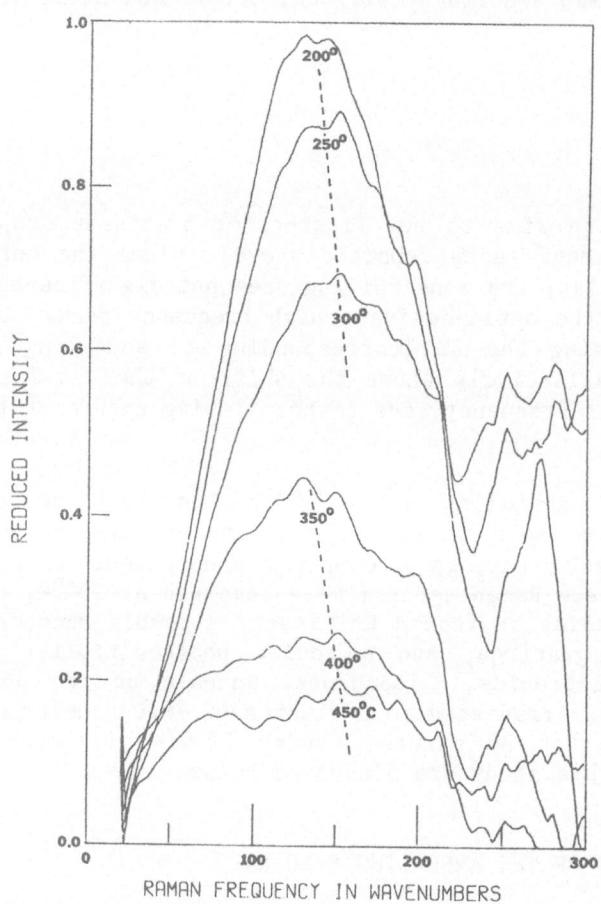

Fig. 2. Computer enhancement of the low frequency damage peak,
        showing not only a shifting behavior toward higher
        frequency, but the appearance of new resonances after the
        350°C and 450°C anneal stages.

be noted that all data points seem to lie on a curve showing a pronounced break at approximately 250°C. Most evident in the low frequency data for p-Si (open circles), is a peak in the anneal curve at 350°C, due in large part to the new defect mode appearing at that temperature.

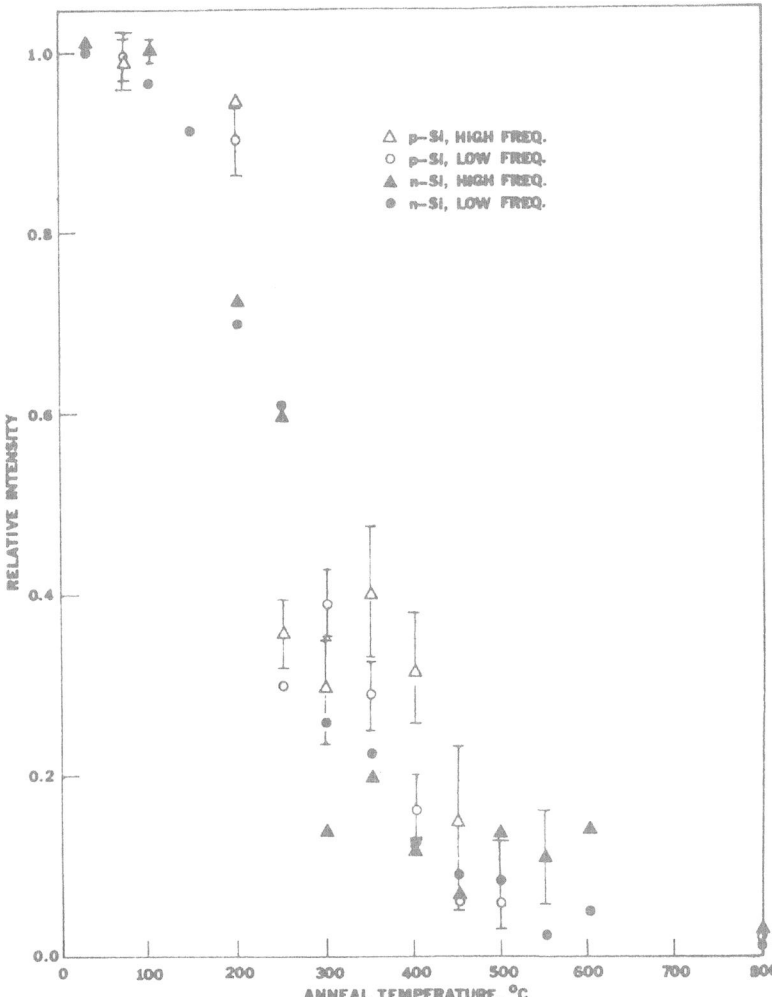

Fig. 3. Normalized intensities of both the high (490 cm^{-1}) and low (120 cm^{-1}) frequency-damage peaks, as a function of anneal temperature. Open triangles and circles represent annealing data for the p-Si sample. Solid figures are data points for the n-Si sample. Note the peak, at approximately 350°C, for the computer-enhanced data (open circles, representing normalized intensity for the low-frequency peak). This peak is suspected to be the result of P-1 growth and decay.

The annealing curve of Fig. 3 was smoothed by averaging data points at each anneal temperature. The resulting curve was then fit to various isochronal annealing schemes, in both the low (30°~300°C) and high (350°~600°C) annealing temperature ranges. In both regimes, the data was found to most closely obey the second-order, isochronal decay law

$$\ln \frac{1}{f_{i+1}} - \frac{1}{f_i} = \ln(\nu_0 N_1 \Delta t) - \frac{E}{k_B}\left(\frac{1}{T_{i+1}}\right) , \qquad (1)$$

where f_i represents the fraction (proportional to Raman intensity) remaining after the ith anneal pulse, at temperture T_i. N_1 is the number fraction of the defect at the initiation of the first anneal. The rate constant, $k_{i+1} = \nu_0 \exp(- E/k_B T_{i+1})$, involves an intrinsic jump rate ν_0, as well as an activation barrier, E.

The low and high temperature regions named above have therefore been shown to obey second-order annealing kinetics. For the temperature range 30°~300°C, $k_1 = 3.6 \times 10^6 \exp(- 5797/T_i)$, corresponding to an activation barrier of 0.50 ± 0.10 eV. For the annealing range from 350°~600°C, $k_2 = 4.0 \times 10^{10} \exp(- 13500/T_i)$, corresponding to an activation barrier of about 1 eV.

Interestingly, complementary resistivity data for the p-Si sample, interpreted as fractional carrier recovery, was likewise determined to be second-order in nature, yielding an activation barrier of 0.65 ± 0.10 eV.

3. DISCUSSION

In the ESR studies carried out by Katz and Hale (1979) on NTD silicon, the P-1 center (5-vacancy) was observed to be the dominant defect in the sample after the 350°C annealing stage. In the present work, a distinct peak is seen in the anneal curve at this temperature, corresponding to this particular defect (Fig. 4). The similarity in the history of the samples used in this Raman study and those of the ESR study strongly suggests that the peak in the Raman intensity at 300°C (Fig. 3), is due to the P-1 centers. Thus, the resonance observed at 128 cm^{-1} (at 350°C) would represent a normal mode of the 5-vacancy.

As has been suggested by Meese et al. (1980), the P-1 center is not the dominant defect in NTD silicon, until the 300°C anneal stage, where it has been conjectured that the P-1 center forms by the combination of smaller vacancies, such as the divacancy (V-2) and the 4-vacancy (P-3). Beyond 450°C, the P-1 centers are thought to vanish by aggolmeration with other vacancies to form large aggregates.

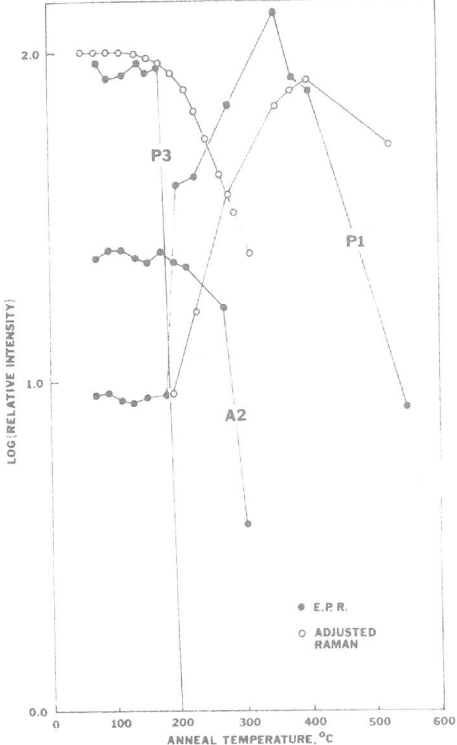

Fig.4.   The annealing data of the Raman study, adjusted (via the
         assumed model) for 10-minute isochronal anneals (open
         circles).   Solid circles indicate annealing curves of
         specific point defects in silicon, as identified in ESR
         work.   Note the strong correlation of the adjusted Raman
         data with both the P-3 and A-2 anneal curves, in the
         temperature range from 30°C to 300°C.   Also observe the
         close correspondence, both in intensity and peak
         temperature of the P-1 anneal curve and the adjusted
         Raman intensity, in the temperature range from 300°C to
         400°C.

     With the P-1 center having both source and sink, a secular
decay scheme may be imagined, in which the rate of P-1 production
is given by

$$\frac{dN}{dt} = k_1 (N_0 - N)^2 - k_2 N^2 \ , \tag{2}$$

where we have incorporated the second-order nature of the anneal,
as established in the previous section, assuming that both source

and sink terms depend upon the square of the defect concentration. Here N and N_o are the concentrations of the P-1 and P-3 centers and k_1 and k_2 are the rate constants. One possible explanation for the source term, would be the combination of P-3 and V-2 centers, known to exist as the majority defects in nearly equal concentrations of about $N = 5 \times 10^{17}$ cm^{-3}, just following room temperature irradiation (Meese, 1980).

By solving Eq. 2, using the experimentally obtained rate constant k_1 and k_2, it was possible to adjust the Raman data for any arbitrary annealing history. By regenerating the Raman data for 10 minute isochronal anneals (as opposed to the original 30 minute anneals) we could then directly compare the ESR and Raman results. Matching the adjusted Raman intensity to the ESR intensity at 200^oC, we obtain the comparative plot shown in Fig. 4. We see qualitative agreement in peak position as well as in relative intensity, corroborating the point of view that the resonance at 128 cm^{-1} is due to the P-1 center. It is to be noted that use of only the computer enhanced data for the low frequency peak in establishing k_1 and k_2 (as opposed to the average of all the data), brings both adjusted Raman peaks into nearly perfect coincidence with the ESR peak at 350^oC.

Similarly, both ESR and Raman data may be matched at 300^oC, to show a strong correlation between the adjusted Raman intensity and the annealing of both P-3 and A-2 centers.

In view of the relatively low activation energy in the low temperature regime (0.50 eV) and the low attempt frequency obtained (10^6 sec^{-1}), it is possible that the mobilization of the defects which ultimately combine to form the P-1 centers, depends upon the ionization of a carrier from these defects. That an activation energy within the experimental precision of this 0.50 eV barrier has been obtained from the resistivity data, strengthens this very point of view.

The overall shifting behavior of the low-frequency peak may be explained by noting that the very asymmetric modes represented by this peak (Thaler, 1982) samples much more of the bond-bending force constants of the surrounding lattice, than its high frequency counterpart. From Brodsky (1975), we learn that the fundamental perturbation in the damaged lattice manifests itself in the form of altered bond-bending force constants. Hence the low-frequency peak should be more sensitive to the surrounding lattice damage and should shift upward in frequency as the lattice environment hardens through annealing.

4. REFERENCES

M.H. Brodsky, Light Scattering in Solids, ed. M. Cardona (Springer-Verlag), New York, 1975, p. 208.

M. Chandrasekhar, H.R. Chandrasekhar, J.M. Meese, and S.L. Thaler, in Defects and Radiation Effects in Semiconductors, (The Institute of Physics, Bristol and London, 1980), ed. R.R. Hasiguti, p. 205.

M. Chandrasekhar, H.R. Chandrasekhar and J.M. Meese, Solid State Comm., 38, 1113 (1981).

L. Katz and E.B. Hale, Neutron Transmutation Doping in Semeconductors, (Plenum, New York, 1979), ed. J.M. Meese, p. 281.

J.M. Meese, Ph.D. Thesis, Purdue University, 1970 (unpublished).

J.M. Meese, M. Chandrasekhar, D.L. Cowan, S.L. Chang, H. Yousif, H.R. Chandrasekhar and P.M. McGrail, to be published in the proceedings of Neutron Transmutation Doping in Semiconductors, (Copenhagen, 1980).

S.L. Thaler, Ph.D. Thesis, University of Missouri-Columbia, 1982 (unpublished).

5. ACKNOWLEDGMENTS

This research was funded by a grant from a special fund provided by the Alumni Development Board of the University of Missouri-Columbia.

ANNEALING STUDY OF NTD SILICON DOPED WITH BORON

Jeffrey B. Watson* and B. C. Covington

Sam Houston State University
Huntsville, Texas 77341

ABSTRACT

Two silicon samples containing approximately 1×10^{15} B-cm^{-3} have been transmutation doped to phosphorous concentrations of 1×10^{13} and 1×10^{15} P-cm^{-3}. For each sample, the infrared absorption properties in the spectral region 4000 to 200 cm^{-1} have been studied as a function of annealing temperature. The samples were isochronally annealed from 200°C to 1000°C in steps of 100°C. After each annealing step, absorption data were taken at room temperature and 10 K. For each annealing step, the optical activation of the phosphorous and boron, as well as the removal or formation of defect lines, is reported.

INTRODUCTION

Since the beginning of neutron transmutation doping (NTD) of silicon, interest in the effects of annealing on radiation-produced defect centers has steadily grown. As interest in the subject has grown, so too has research, yet there have been few publications[1,2] on the effects of annealing temperature on the optical properties of NTD silicon which contains significant concentrations ($\geq 10^{15}$ cm^{-3}) of boron. This paper presents the initial results of a research effort to obtain the optical properties as a function of annealing temperature and irradiation time of NTD silicon which contains approximately 10^{15} B-cm^{-3}.

---

*Present address, 339 Sykes Circle, Ridgecrest, Ca. 93555

EXPERIMENTAL

Sample Material

The Czochralski-grown boron-doped silicon used in this research was donated by the Air Force Materials Laboratory. Resistivity values of 8.85 to 10 Ω-cm indicate a boron concentration of approximately 10^{15} cm^{-3}. The oxygen concentration is of the order of 10^{18} cm^{-3}. Five samples each measuring 13 x 13 x 2 mm were cut from adjacent positions in the boule and then irradiated in the Missouri University Research Reactor to nominal phosphorous concentrations of 1 x 10^{13}, 5 x 10^{13}, 1 x 10^{14}, 5 x 10^{14}, and 1 x 10^{15} cm^{-3}. An unirradiated sample to be used as a reference was cut from the same boule as the irradiated samples. The results reported here are for the most- and least-irradiated samples.

Infrared Scanning

A Perkin-Elmer double-beam infrared spectrophotometer was used to obtain room temperature and 10 K absorption data in the spectral region 4000-200 cm^{-1}. The instrumental resolution at 500 cm^{-1} was approximately 3 cm^{-1}. White light was incident on the sample at all times during the scan to insure that all impurities were ionized.

Cryogenics

The low-temperature scans were taken with the sample mounted to the cold finger of a Cryodyne closed-cycle refrigerator. Copper impregnated grease was used to keep the sample in thermal contact with the cold finger and to avoid stressing the sample. Cesium iodide windows were used on the vacuum shroud of the cold finger.

Annealing

After obtaining the initial room temperature and 10 K spectral scans, each sample was annealed at 200°C for fifteen minutes. The samples were annealed in a fused-quartz boat and tube that was purged with argon gas. The absorption spectrum of each sample was then retaken. The procedure was continued in isochronal annealing steps of 100°C until 900°C was reached. A final anneal at 1000°C for 30 minutes was made to check for any additional changes in the optical properties resulting from a higher temperature and more prolonged annealing.

RESULTS

The transmission spectrum at 10 K for the unannealed reference sample is shown in Figure 1. Only the spectral region 1175 cm^{-1} to 200 cm^{-1} is shown because most of the significant spectral changes

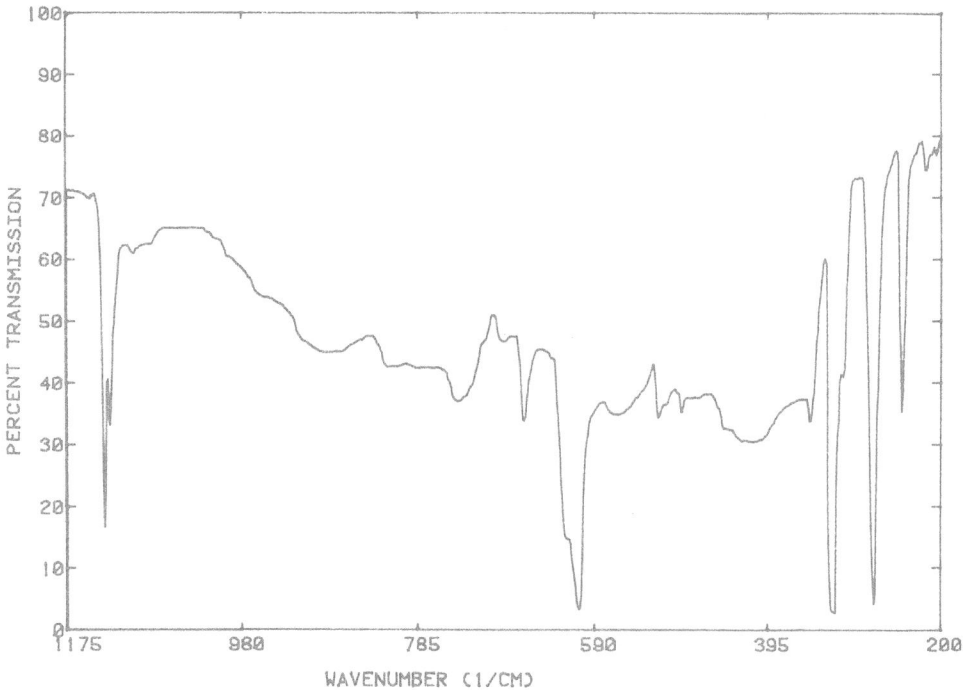

Fig. 1.   Spectrum of the unannealed reference sample at 10 K.

occurred in this region. The regions that will be of particular
interest with respect to the scans for the irradiated samples are
(1) the oxygen lines around 1137 cm^{-1} (2) the $P_{1/2}$ boron lines at
668 and 692 cm^{-1} (3) the $P_{3/2}$ boron lines in the region 245 to
350 cm^{-1} and (4) the region around 440 cm^{-1}.

Since no significant changes occurred in the spectra of any
NTD sample after the 200 and 300°C anneals, only spectra taken after
higher annealing temperatures will be shown. The spectrum of the
least-irradiated sample (1 x 10^{13} P-cm^{-3}) after the 400°C anneal is
shown in Figure 2. It should be noted that the $P_{1/2}$ and $P_{3/2}$ boron
lines are not present and that two unidentified lines appear in the
200 - 250 cm^{-1} region. After the 500°C anneal, some of the $P_{3/2}$ and
$P_{1/2}$ boron lines began to reappear and continued to increase in
intensity with each annealing step. The 900°C anneal and the 1000°C
anneal for the least-doped sample are shown in Figures 3 and 4,
respectively. These spectra when compared to Figure 1 indicate
that even though some of the boron $P_{1/2}$ and $P_{3/2}$ lines have been
reactivated that a significant amount of the boron is still not
optically active. The inability to accurately identify any phosphorous lines in the least-doped sample's spectra can be attributed
to two possibilities. Either the phosphorous is not optically
active or, due to the low phosphorous concentration (1 x 10^{13} cm^{-3}),

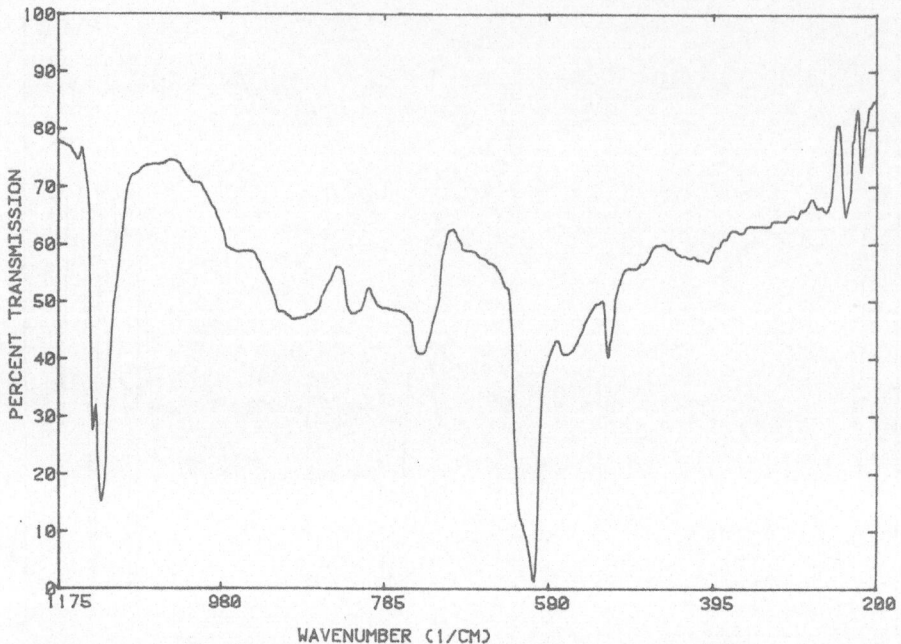

Fig. 2. Least-irradiated sample at 10 K after 400°C anneal.

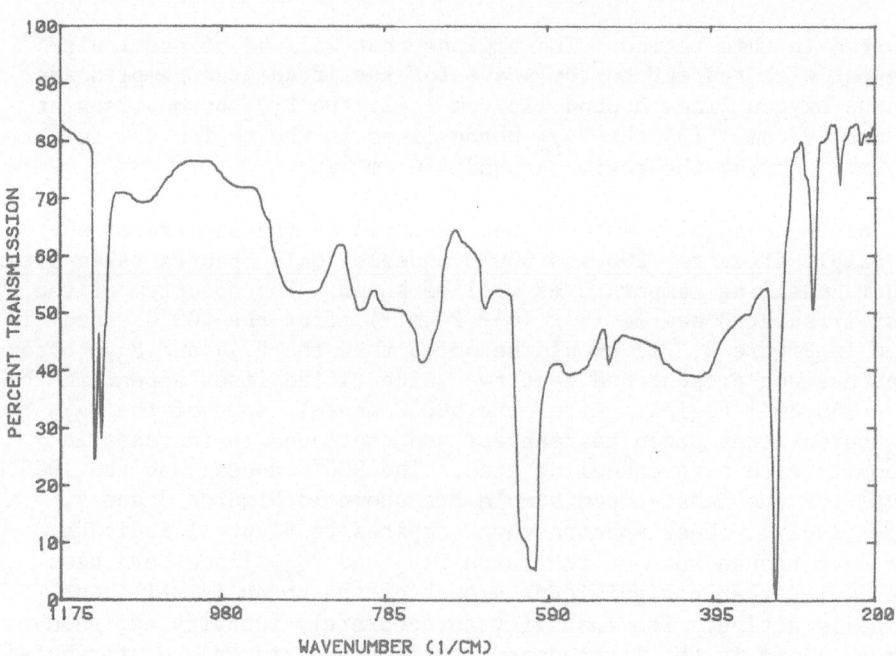

Fig. 3. Least-irradiated sample at 10 K after 900°C anneal.

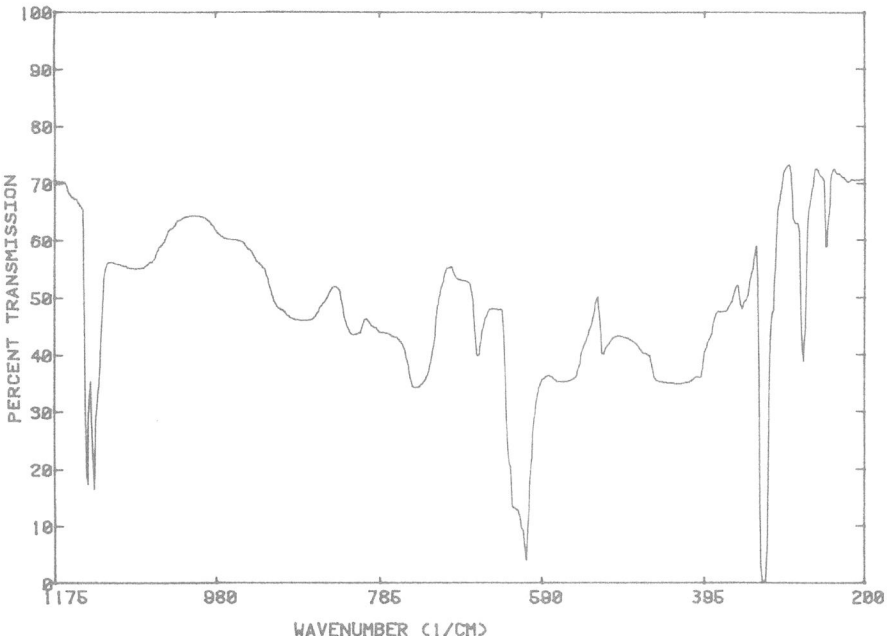

Fig. 4. Least-irradiated sample at 10 K after 1000°C anneal.

the boron spectrum completely covers the phosphorous spectrum.
From data obtained for the most irradiated sample, it appears that
the phosphorous is not optically active. The spectra after the
900 and 1000°C anneals for the most irradiated sample are shown in
Figures 5 and 6, respectively. These spectra show that there have
been some recovery of the boron $P_{1/2}$ and $P_{3/2}$ lines, but the number
of lines and the individual line strength is obviously less than the
reference sample. In addition, the large broad bands at approximately 1070 cm^{-1} and 440 cm^{-1} were not removed by the high
temperature anneal. These bands were present before the 200°C
anneal and continued to gradually increase in intensity with each
annealing step. They were observed in both sample's spectra but
are much more intense in the most irradiated sample. There is some
question as to the exact cause of these bands. Infrared absorption
spectra in the spectral region 1000 to 1400 cm^{-1} for stressed NTD
silicon have been reported by Corelli and Corbett.[3] They report
several rather sharp lines in the region around 1070 cm^{-1}, which
they classify as Higher Order Bands (HOB). All of the HOB reported
in their work were removed by annealing at approximately 600°C.
Absorption data for electron irradiated silicon reported by Newman[2]
shows at 441 cm^{-1} an absorption peak which is attributed to a
resonant mode of substitutional phosphorous. This peak is considerably sharper than the broad band shown in our spectra.

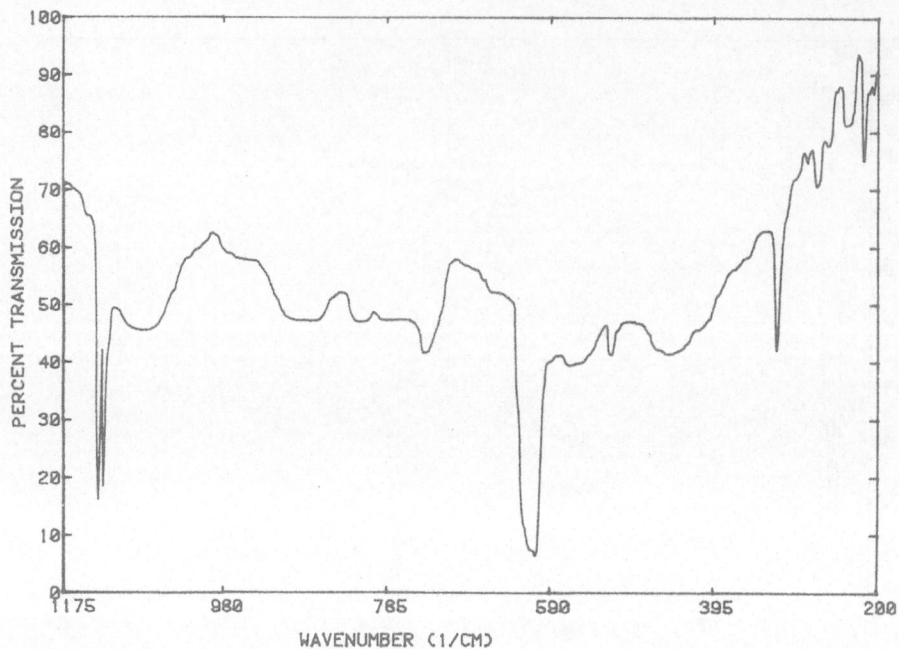

Fig. 5.   Most-irradiated sample at 10 K after 900°C anneal.

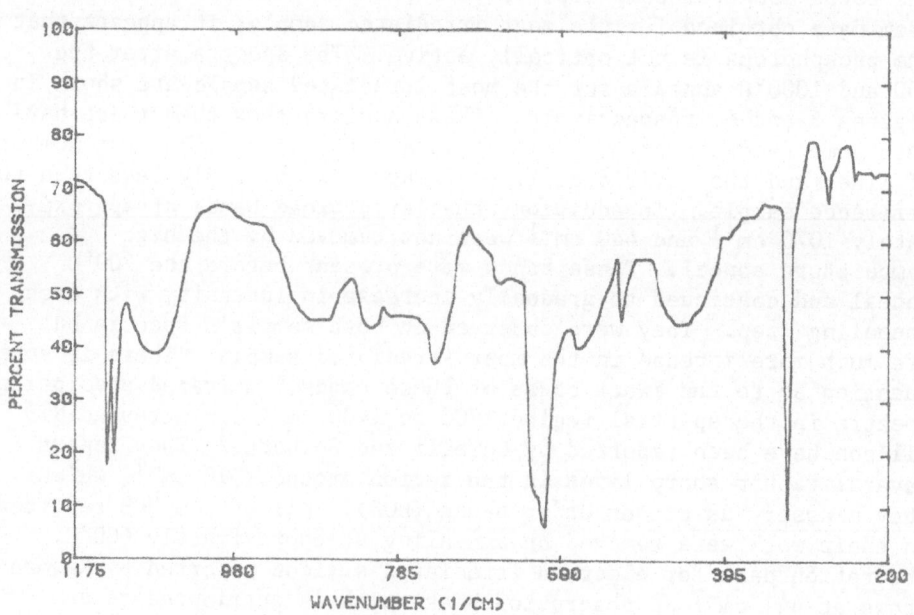

Fig. 6.   Most-irradiated sample at 10 K after 1000°C anneal.

SUMMARY AND CONCLUSIONS

We have reported the initial findings of an infrared absorption study of NTD silicon which contains approximately 1×10^{15} $B\text{-cm}^{-3}$. The spectra for two samples transmutation doped to phosphorous concentrations of 1×10^{13} cm^{-3} and 1×10^{15} cm^{-3} are given as a function of annealing temperature, which ranges from 200 to 1000°C After a 1000°C anneal, we were unable to optically activate a detectable amount of phosphorous in either sample. After the 500°C anneal, three of the boron $P_{3/2}$ lines in both samples were activated to a detectable intensity. These $P_{3/2}$ lines gained in intensity with each anneal up to the 1000°C anneal. The $P_{1/2}$ line at 686 cm^{-1} was activated in the least-doped sample after the 500°C anneal and gained intensity with each additional annealing step. For the most-irradiated sample, the 686 cm^{-1} line was detectable only after the 1000°C anneal. After the 1000°C anneal, the 692 cm^{-1} $P_{1/2}$ line for the least-doped sample remained extremely weak. The 692 cm^{-1} line was never activated in the most-irradiated sample.

Two unidentified absorption bands at 1070 cm^{-1} and 440 cm^{-1} are reported. These bands which were present in both samples increased in intensity as the annealing temperature was increased.

ACKNOWLEDGEMENTS

The authors wish to thank Sam Houston State University, The Air Force Office of Scientific Research, and The Research Corporation for their financial assistance. We also wish to thank The Air Force Materials Laboratory for the samples and Dr. Robert Spry for his many helpful discussions.

REFERENCES

1. R. N. Thomas, T. T. Braggins, H. M. Hobgood, and W. J. Takei, J. Appl. Phys. 49 (5), 2811 (1978).
2. R. C. Newman, "Neutron-Transmutation-Doped Silicon," Plenum, New York (1981), ed., Jens Guldberg, p. 83.
3. John C. Corelli and James W. Corbett, "Neutron-Transmutation-Doped Silicon," Plenum, New York (1981), ed. Jens Guldberg, p. 35.

ANNEALING EFFECTS OF NTD SILICON ON HIGH-POWER DEVICES

C.K. Chu And J.E. Johnson

Westinghouse Electric Corporation
Semiconductor Division
Youngwood, PA 15697

ABSTRACT

Neutron-transmutation doped (NTD) silicon has become very important for the manufacture of high-power devices with almost all power-device manufacturers using it. It is well known that NTD silicon yields a higher reverse blocking-voltage distribution than float zoned silicon. The forward conduction voltage drop of a power device is also a very important parameter. This paper shows that by using higher-lifetime NTD starting material, the forward-voltage drop distribution can be lowered. Annealing the unannealed NTD silicon during device fabrication may increase the minority-carrier lifetime of the silicon which would then improve the forward-voltage-drop distribution.

INTRODUCTION

Neutron transmutation doped (NTD) silicon has been used for the manufacture of high-power devices for almost ten years. It is well known that NTD silicon yields a higher blocking-voltage distribution than float-zoned silicon.[1] [2] The forward conduction voltage drop of power devices is also a very important parameter. This paper shows that the forward-voltage distribution can be lowered by using higher lifetime NTD starting material. When purchased in quantity, the photoconductive-decay lifetime of NTD silicon is between 200 to 500 μs which is much lower than float-zoned material, which can go as high as 1000 to 2000 μs. The low lifetime of NTD silicon could be caused by:[3]

1. Impurity contamination during silicon cleaning
    before low-temperature annealing.

2. Original silicon crystal fabrication procedures.

3. Type of atomic reactor which is used for neutron
    transmutation doping.

Normally we feel the first item is the major cause of the
low-lifetime NTD silicon. The general applied annealing process
for NTD silicon is done at low temperature, under 1000°C and for
a short time.[4] Almost all NTD silicon suppliers do the low
temperature annealing before they ship the silicon crystals to the
device manufacturers. High-power device manufacturers process NTD
silicon through very high temperature diffusion processes between
1200°C and 1250°C for longer periods of time. In order to study
the high-temperature annealing effects on NTD silicon minority-
carrier lifetime and device characteristics, several annealed and
unannealed NTD silicon crystals were obtained. The unannealed
silicon could then be tested for the effectiveness of annealing
during normal high power device diffusion fabrication procedures
and compared to the annealed material.

EXPERIMENTAL PROCEDURES

For the study of NTD-silicon annealing effects on high-power
devices, we chose 4000 volts blocking, two-inch diameter thyristors
as the experimental vehicle. The design resistivity of the silicon
for these thyristors is about 220 Ω-cm with minimum lifetime of
100 μs. The unannealed silicon crystal had the same specification
as the annealed material. To verify the resistibity of the unannealed silicon crystal, the silicon supplier cut one slice from
each crystal then diffused phosphorous into the slice in order to
check the bulk resistivity of the slice against our specification.
We obtained three crystals of each kind with minimum length of
about four inches. All crystals met our inspection procedures,
and were sliced to the proper thickness with standard Internal
Diameter saws. The silicon wafers were cleaned thoroughly with
our normal procedures, and the two groups of annealed and unannealed silicon wafers were properly identified and processed side
by side through high-power thyristor fabrication procedures in
our Engineering Laboratory and production line. There were 150
wafers from the annealed silicon crystals and the same amount from
the unannealed silicon crystals.

We divided these wafers equally into three different experimental runs. Each run contained 50 annealed wafers and another 50
unannealed wafers. As mentioned before, each run was processed in
the same way and used the same process equipment in order to

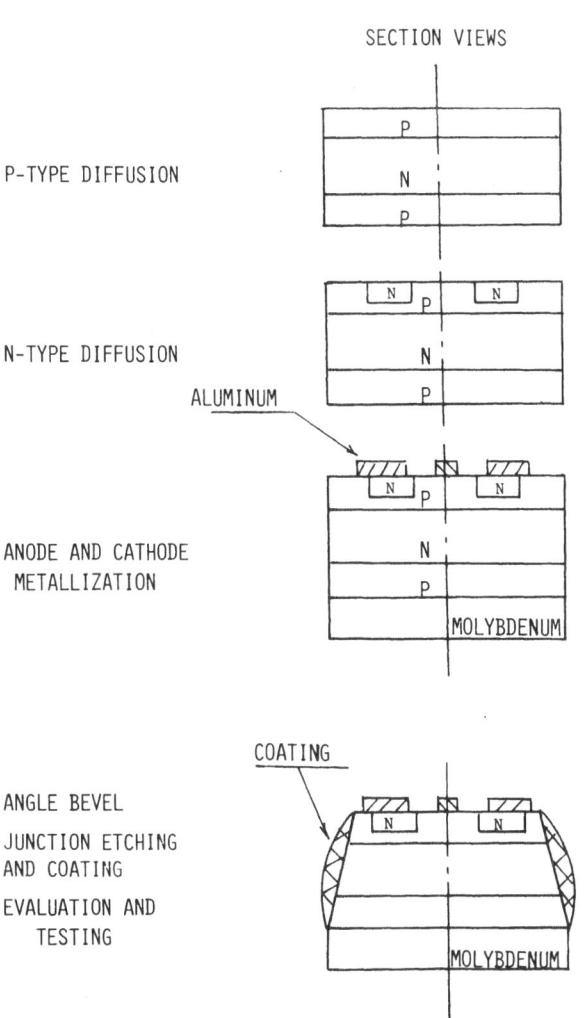

Figure 1. Fabrication Procedure For High Power Thyristors

TABLE 1

ELECTRICAL CHARACTERISTICS OF 50 mm
DIAMETER HIGH-POWER THYRISTORS
(TYPICAL VALUES)

PARAMETERS	ANNEALED BY SILICON MANUFACTURER	ANNEALED DURING DIFFUSION
V_{DRM} (V) @ 25°C 10 MA	4200 Volts	4200 Volts
V_{DRM} (V) @ 125°C 150 MA	4000 Volts	4000 Volts
V_{RRM} (V) @ 25°C 10 MA	4200 Volts	4200 Volts
V_{RRM} (V) @ 125°C 150 MA	4000 Volts	4000 Volts
V_{TM} (V) @ 1500A @ 25°C	1.75 Volts	1.75 Volts

minimize the process variation during device fabrication. The general high-power thyristor element fabrication procedures are listed in Figure 1. The finished elements were properly identified and measured for electrical characteristics. The most important parameters for high-power thyristors are forward and reverse-blocking voltages, the forward-voltage drop, the gate and switching characteristics. Table 1 lists the comparison of the measured electrical characteristics of the thyristors. During processes of these three runs, one run was rejected due to malfunction of a diffusion furnace. The other two runs produced acceptable thyristors with the characteristics shown in Table 1.

It is interesting to note that the unannealed wafers gave a 10% better yield in blocking voltage and forward-voltage drop distribution. Since there were only 100 wafers from each group, more detailed study is necessary to verify the reproducibility of these results.

SUMMARY

In this paper we have demonstrated that the device manufacturer could gain small amount of yield distribution both in forward-voltage drop and blocking voltage distribution by using unannealed NTD silicon crystal.

This improvement could be due to special cleaning before diffusion and high temperature annealing during diffusion.

REFERENCES

1. C. K. Chu and J. E. Johnson "NTD Silicon on High Power Devices," Page 53, Neutron Transmutation Doping in Semiconductor 1979 Plenum Press, New York
2. K. Plazeder and K. Loch IEEE Transactions, Electron Devices, ED-23, No. 8, August, 1976
3. Olof Malmros "The Minority Carrier Lifetime of Neutron Doped Silicon," Neutron Transmutation Doping in Semiconductor 1979 Plenum Press, New York, Page 249
4. Paul J. Glairon and J. M. Meese "Isochronal Annealing of Resistivity in Flot Zone and Czochralski NTD Silicon," Neutron Transmutation Doping in Semiconductor 1979 Plenum Press, New York, Page 291

STUDY OF ANNEALING BEHAVIOR AND NEW DONOR FORMATION IN NEUTRON

TRANSMUTATION DOPED SILICON GROWN IN A HYDROGEN ATMOSPHERE

Wang Zhengyuan and Lin Lanying

Institute of Semiconductors, Chinese Academy of Sciences
Beijing, People's Republic of China

ABSTRACT

This paper shows that at annealing temperatures in the range
(350-600°C), NTD float-zoned silicon (FZ-Si) grown in a hydrogen
atmosphere and NTD grown in an argon atmosphere have very different annealing behavior. This is demonstrated by measurements of
resistivity and Hall effect. The experimental results provide
evidence for a new kind of donor, "hydrogen-defects" complex, in
the NTD FZ-Si grown in hydrogen. The behavior of this new donor
center is studied by means of electrical, optical, and DLTS measurements. After 400-450°C annealing the concentration of this
donor reaches the maximum. In addition, three new hydrogen-
related energy levels, (26 ± 1) meV, (37 ± 1) meV, and (265 ± 5)
meV, have been observed. A shallow donor energy level (26 meV) is
produced by 450°C annealing and two shallow donor energy levels
(26 and 37 meV) are produced by 500°C annealing. The present experiments verify that the formation of this new donor center depends on the existence of the Si-H bond in the silicon and that
these new donors are completely different from the well-known
"oxygen donor" formed in CZ-Si at 450°C. These observations provide new evidence for the electrical activity of hydrogen in silicon.

1. INTRODUCTION

There have been a large number of investigations on NTD
float-zoned silicon grown in argon or vacuum and on Czochralski
silicon. However, there are no detailed reports concerning the
annealing behavior of NTD FZ-Si single crystal material grown in a
hydrogen atmosphere.

On the other hand, based on a similarity between the atomic structure of hydrogen and that of lithium, there has been considerable speculation about the donor behavior of hydrogen in silicon. Hydrogen has been used to suppress the formation of swirl defects in FZ silicon, and it was observed that the behavior of amorphous silicon was very much affected by hydrogen. It was also found that the atomic hydrogen can passivate point defects in silicon and this technique has been applied to device fabrication. These considerations motivated studies of the behavior of hydrogen in silicon. In 1971, Zohta et al.[1] first reported the experimental results on the electrical activity of hydrogen in silicon. They made use of proton-bombarded silicon and measured shallow donor levels (26 ± 1) meV below the conduction band after an annealing at about 300°C. They concluded that this was due to atomic hydrogen located at a specified site on the silicon lattice. However, Picraux et al.[2] pointed out in 1978 that "only a minimal understanding of the electronic structure of hydrogen-related defect centers in crystalline silicon currently exists."

This paper shows that during an annealing in the range of 350-650°C, NTD silicon grown in a hydrogen atmosphere exhibits a new donor. We conclude that this new donor is associated with hydrogen in silicon, and the following properties of this new donor center were investigated: (1) its activation energy; (2) its dependence on the Si-H bond in silicon; and (3) its differences from the already known oxygen donor formed in CZ crystal silicon after annealing at 450°C. NTD FZ-Si grown in a hydrogen atmosphere has proven to be a very effective material for studies of the behavior of hydrogen in silicon.

2. EXPERIMENTS

2.1 Samples

An ingot of single-crystal silicon grown by float-zone techniques in a hydrogen atmosphere was selected that had the following parameters: N-type, $\rho > 1,000$ $\Omega \cdot$cm, $\tau \approx 1,000$ μs, $[O] < 5 \times 10^{15}$ cm^{-3}, and $[C] < 2 \times 10^{16}$ cm^{-3}. It was cut into three segments: segments A and B were neutron-doped in heavy and light water reactors, respectively, and segment C was not NTD-doped but kept as a control specimen. These three segments were given isochronal anneals in nitrogen for 30 minute intervals from room temperature to 900°C with a temperature step of 50°C. Another segment called D was cut off from segment C before the isochronal annealing and preannealed at 650°C for 0.5 hours and NTD-doped in the light water reactor in the same way as segment B. The irradiation conditions for these four samples are shown in Table I.

Table I. Irradiation Conditions for the Samples A, B, C, and D.

Sample	Doping	$\Phi_{th}t$	R	ρ_t	L	Remarks
A	NTD in H.R.	9.7×10^{16}	100	190	43. 0	
B	NTD in L.R.	1.8×10^{17}	5	150	43. 6	
C	not NTD Doped	0		1100	56. 6	
D	NTD in L.R.	1.8×10^{17}	5	150	25. 0	Preannealed at 650°C for 0.5 hours before NTD doping

H.R. --- heavy water reactor   L.R. --- light water reactor

$\phi_{th}t$ --- the thermal neutron fluence in neutron/cm^2

R --- the ratio of thermal neutron flux and fast (>100 eV) neutron flux

ρ_t --- target resistivity in ohm-cm

L --- sample length in mm

2.2 Measurements

Resistivity, carrier concentration, Hall mobility, and infrared absorption spectra (4–5.5 µm wavelength) were measured for the samples annealed at the different temperatures; the activation energy of the donor or acceptor centers was also measured for some samples annealed at corresponding temperatures. In addition, DLTS spectra of a selected number of samples were also measured.

3. RESULTS

The variation of resistivity and carrier concentration for samples A and B with annealing temperature are shown, respectively, in Fig. 1 and Fig. 2. For sample B, the changes of $nT^{-3/2}$ or $pT^{-3/2}$ vs. $\frac{1000}{T}$ are shown in Fig. 3 for selected measurement temperature ranges. For comparison's sake, the corresponding curves for NTD silicon FZ grown in argon are also shown in the figure. The apparent activation energies corresponding to the slopes of these curves are given in the figure caption.

Figure 4 summarizes the relationship between the relative resistivity of the four samples A, B, C, and D and the isochronal annealing temperature. Notice that the behavior of each sample is different.

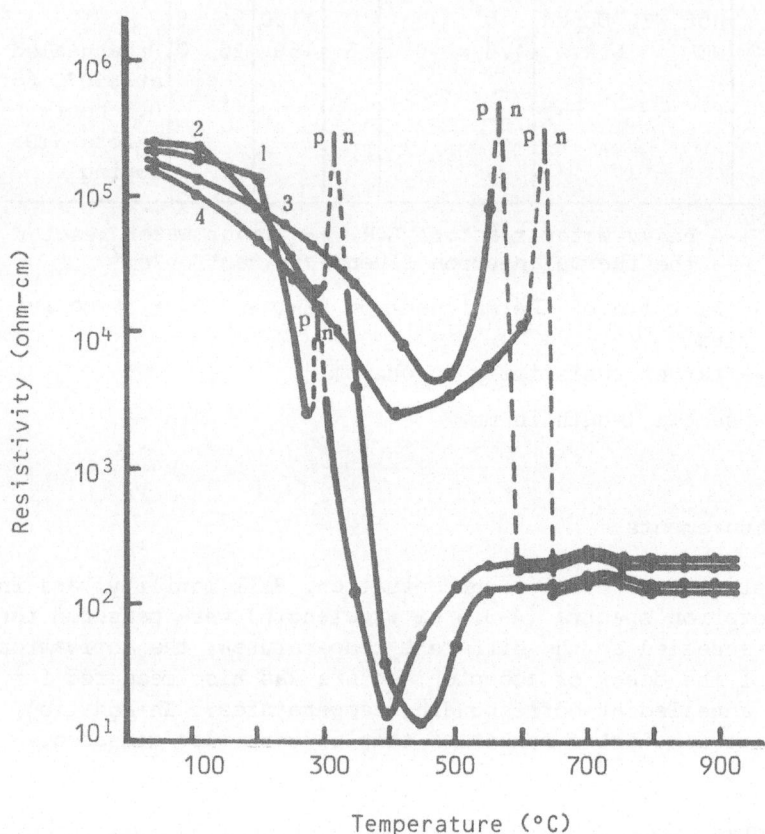

Fig. 1. Resistivity of NTD silicon vs. isochronal annealing temperature.

1. FZ in H_2, NTD in a heavy-water reactor (sample A).
2. FZ in H_2, NTD in a light-water reactor (sample B).
3. FZ in Ar, NTD in a heavy-water reactor.
4. FZ in Ar, NTD in a light-water reactor.

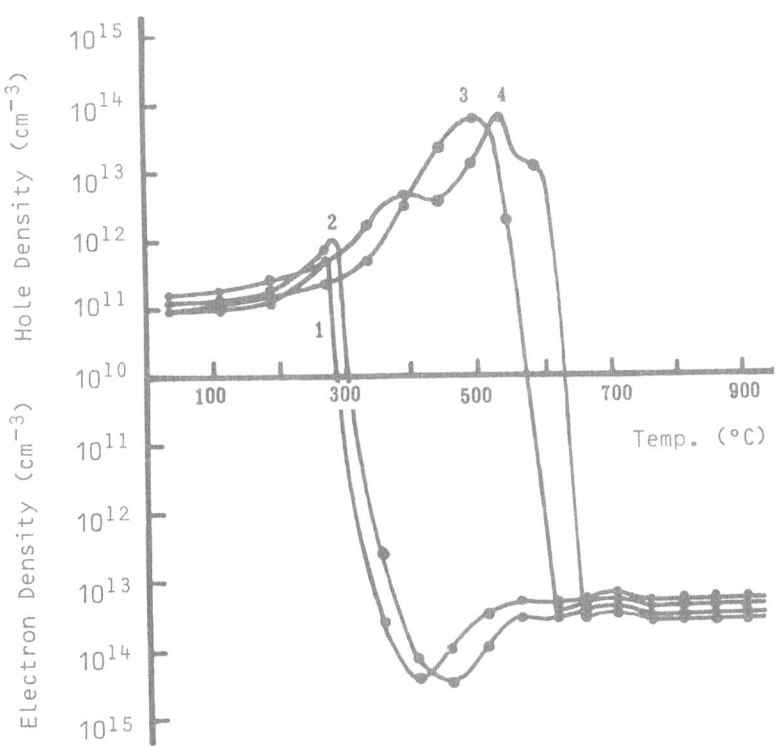

Fig. 2. Carrier concentration of NTD silicon vs. isochronal annealing temperature.

1. FZ in H_2, NTD in a heavy-water reactor (sample A).
2. FZ in H_2, NTD in a light-water reactor (sample B).
3. FZ in Ar, NTD in a heavy-water reactor.
4. FZ in Ar, NTD in a light-water reactor.

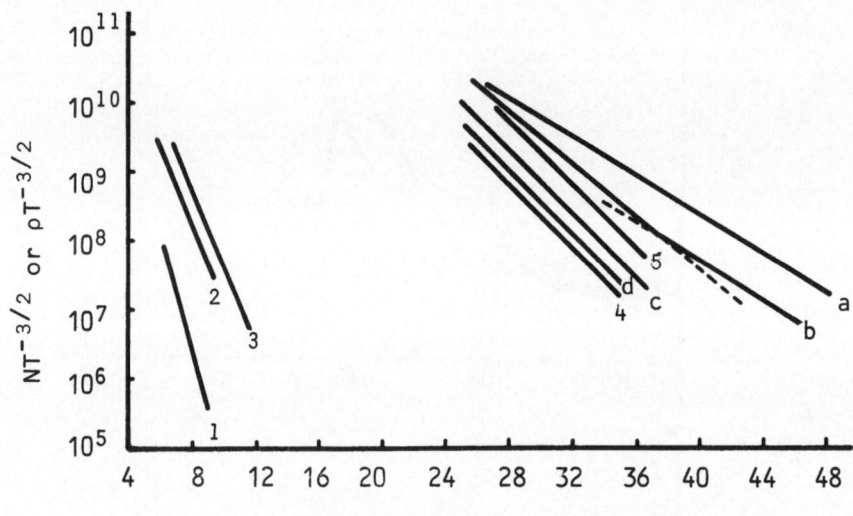

Fig. 3. $pT^{-3/2}$ (or $nT^{-3/2}$) vs. $\dfrac{1000}{T}$, for NTD silicon after isochronal annealing to various temperatures for 30 min each.

a. FZ in H_2, 450°C, n-type, $\Delta E = 27$ meV;
b. FZ in H_2, 500°C, n-type, $\Delta E_1 = 27$ meV, ΔE_2, $= 37.2$ meV;
c. FZ in H_2, 750°C, n-type, $\Delta E = 45.2$ meV;
d. FZ in H_2, 1150°C, n-type, $\Delta E = 45.1$ meV.

1. FZ in Ar, 450°C, p-type, $\Delta E = 0.17$ eV;
2. FZ in Ar, 500°C, p-type, $\Delta E = 0.11$ eV;
3. FZ in Ar, 550°C, p-type, $\Delta E = 0.115$ eV;
4. FZ in Ar, 650°C, n-type, $\Delta E = 46$ meV;
5. FZ in Ar, 850°C, n-type, $\Delta E = 44$ meV.

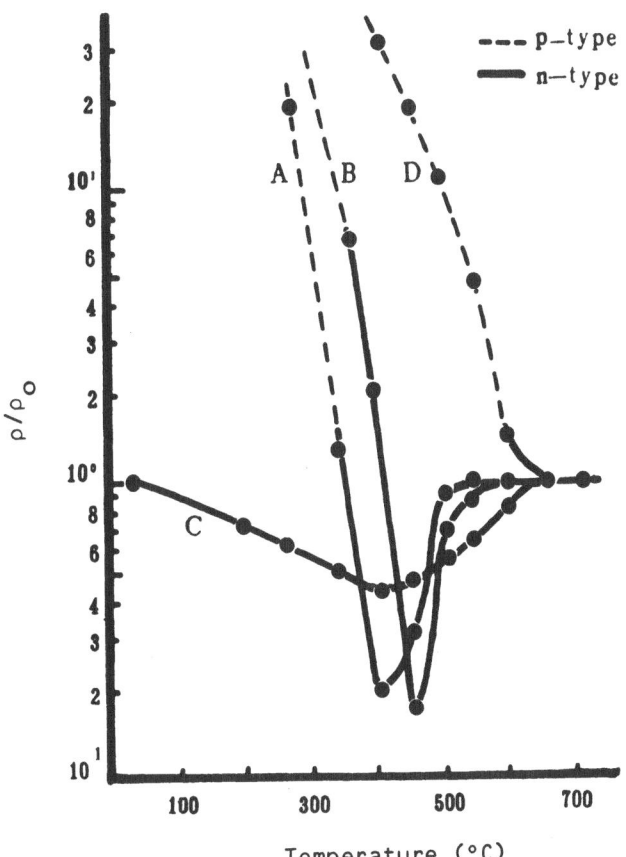

Fig. 4. Relative resistivity for samples A, B, C, and D vs. the isochronal annealing temperature.

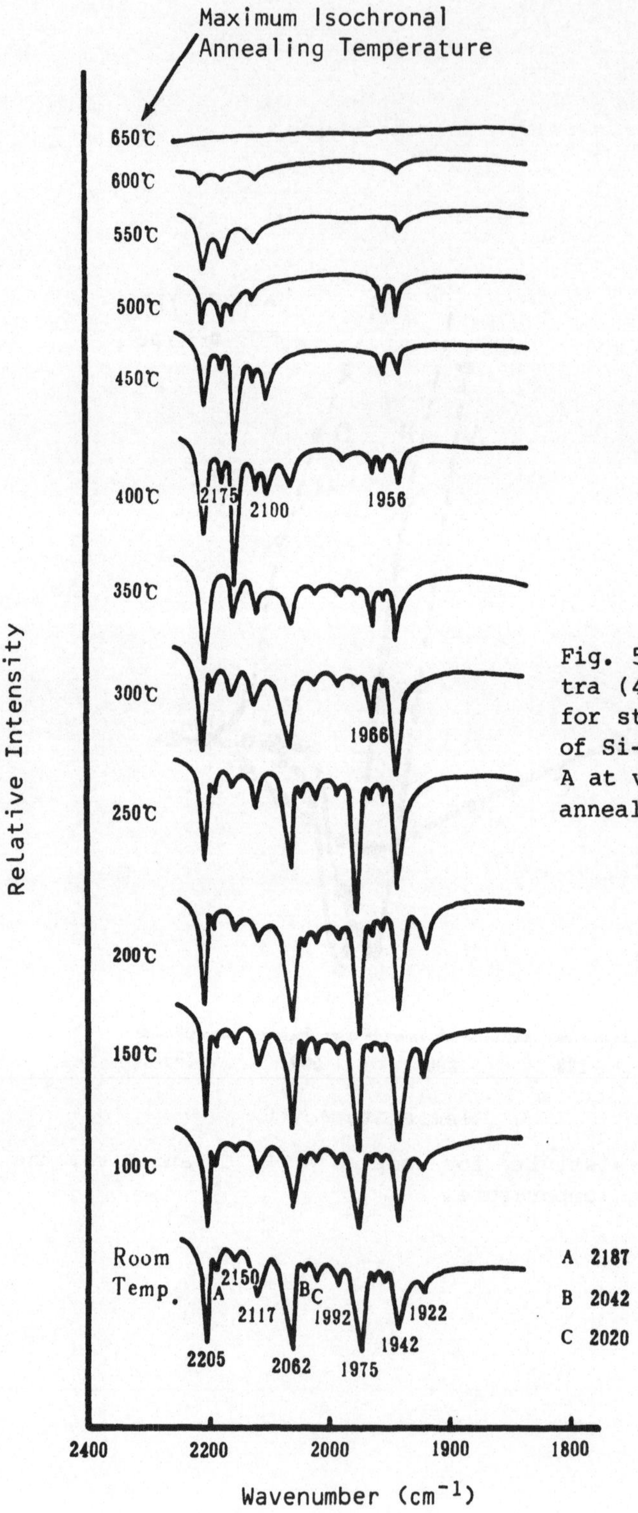

Maximum Isochronal
Annealing Temperature

650℃

600℃

550℃

500℃

450℃

400℃   2175   2100   1956

350℃

300℃   1966

250℃

200℃

150℃

100℃

Room
Temp.   A 2150   B C
           2117   1992   1922
                        1942
   2205   2062   1975

A 2187

B 2042

C 2020

2400   2200   2000   1900   1800

Wavenumber (cm⁻¹)

Relative Intensity

Fig. 5. Infrared spec-
tra (4-5.5 μm wavelength
for stretching vibration
of Si-H bonds) for sample
A at various isochronal
annealing temperatures.

Figure 5 shows the variation of the infrared absorption spectra of sample A with isochronal annealing temperature. Stein[3] has pointed out that the infrared absorption peak in the 4-5.5 μm range corresponds to the stretching vibration of Si-H bonds. It can be seen from Fig. 5 that all absorption peaks disappear after annealing at 650°C. This means that all the Si-H bonds have been broken. The absorption peaks shown in the infrared spectra of Fig. 5 can be divided into four categories:

(1) The samples A, B, and C have the common Si-H characteristic peaks at 2205, 2187, 2117, 1992, and 1942 cm^{-1}. These common peaks have good thermal stability but vanish after the annealing at 500-650°C. The peaks at 2205, 2117, and 1992 cm^{-1} correspond to Si-H stretching vibration of the hexagonal, tetrahedral, and antibonding interstitial sites[4,5].

(2) The common peaks at 2012, 1975, and 1922 cm^{-1} for samples A and B are assumed to result from neutron irradiation since they are not present in the control sample C.

(3) The peaks at 2045, 1889, 1857, and 1827 cm^{-1} occur only in sample B. They are probably associated with defects produced by irradiation in the light-water reactor (high fast neutron

Table II.   Electron-Trap Parameters of NTD FZ-Si Grown in a Hydrogen Atmosphere.

Peak No.	Sample A after isochronal anneal at 400°C			Sample B after isochronal anneal at 450°C		
	E_t eV	n_t cm^{-3}	σ_t cm^2	E_t eV	n_t cm^{-3}	σ_t cm^2
1	$E_c - 0.13$	4.7×10^{12}	3×10^{-15}	$E_c - 0.13$	3.1×10^{13}	5×10^{-15}
2	$E_c - 0.21$	3.4×10^{12}	1×10^{-14}	$E_c - 0.20$	3.6×10^{13}	4×10^{-15}
3	$E_c - 0.27$	1.2×10^{13}	2×10^{-15}	$E_c - 0.26$	5.1×10^{13}	7×10^{-16}
4	$E_c - 0.30$	1.4×10^{13}	2×10^{-15}	$E_c - 0.28$	5.0×10^{13}	5×10^{-16}
5	$E_c - 0.39$	7.8×10^{12}	8×10^{-16}	$E_c - 0.38$	4.2×10^{13}	1×10^{-15}
6	$E_c - 0.41$	8.0×10^{12}	3×10^{-16}	$E_c - 0.40$	2.6×10^{13}	2×10^{-16}

E_t --- Energy level of defect state, n_t --- defect center state density.

σ_t --- Capture cross section of defect center, E_c = Energy of the conduction band minimum.

flux). They have poor thermal stability and disassociate at 100-250°C.

(4) The peaks at 2150, 2100, 1966, and 1956 cm^{-1} are produced during annealing and are sharply increased at certain temperatures. We are presently interested in the peak at 2150 cm^{-1} for reasons to be discussed. All the above absorption peaks did not occur in sample D.

After the isochronal anneals at 400°C and 450°C, Schottky-barrier test devices were prepared from samples A and B and their DLTS spectra were measured. The results for sample B are shown in Fig. 6. These DLTS curves were analyzed by the method of J. Guldberg[7] and the resulting electron-trap parameters for samples A and B are shown in Table II. The capture cross sections in this table were determined by the intercept of an Arrhenius plot.

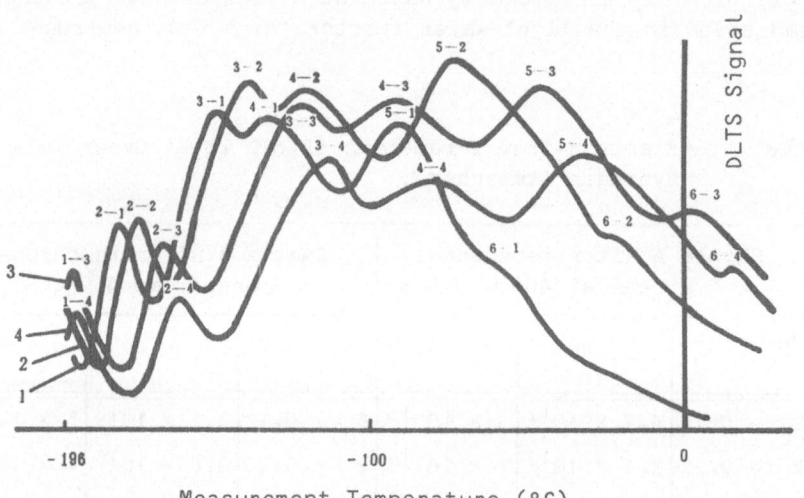

Measurement Temperature (°C)

Fig. 6. DLTS spectra for sample B after 30 min annealing at 450°C. The index i-j identifies the emission peak of the i-th trap center for the j-th reference emission rate. The j values correspond to different emission rate windows as follows:

1 --- 50 s^{-1}; 2 --- 500 s^{-1}; 3 --- 5000 s^{-1}; 4 --- 12500 s^{-1}.

The sample was: N-type; ρ = 14.4 Ω•cm (at room temperature); n_o = 3 × 10^{14} cm^{-3}; trap-filling voltage = 4.5 V; voltage during DLTS transient = 5.0 V; steady-state capacitance = 2.33 pF.

4. DISCUSSION

(1) Annealing behavior of NTD silicon FZ grown in H_2

Because of the high density of electrically active irradiation-induced defects and their motion, disassociation, and transformation into other forms of defects or complexes during annealing, the conduction type of the neutron irradiation silicon will experience changes from P- to N-type with associated resistivity variations of 4 to 5 orders of magnitude.

Figures 1 and 2 show that during the annealing at temperatures below 300°C as well as above 650°C, NTD silicon FZ grown in hydrogen or argon shows similar behavior. However, during annealing between 350-600°C, the behavior in these two cases is quite different. The material grown in argon remains P-type with its acceptor activation energy being, respectively, 0.17 eV (after the annealing at 450°C) and 0.12 eV (after the annealing at 500°C and 550°C), and then changes into N-type at about 600°C. The material grown in hydrogen changes into N-type at about 300°C. Obviously, this is because a compensating donor of even higher concentration appears in this material. Figure 2 shows that after an annealing at about 400°C, the electron concentration of this N-type material is one order of magnitude higher than the concentration of transmuted phosphorus. It follows that the conductivity in this case is not due mainly to the contribution of phosphorus, but from the hydrogen-related new donor.

(2) The behavior of hydrogen-related new donor

a. Electrical properties of the new donor

The variation of resistivity and carrier concentration with the annealing temperature presented in Figs. 1 and 2 show that this donor center begins playing a part after annealing at about 300°C, and that its concentration reaches a maximum after annealing at about 400°C and that it vanishes after annealing at about 550°C. According to our measurements (Fig. 3), the activation energy of this donor is as follows: $\Delta E = (26 \pm 1)$ meV after annealing at 450°C and exhibits two energy levels after annealing at 500°C: $\Delta E_1 = (26 \pm 1)$ meV and $\Delta E_2 = (37 \pm 1)$ meV. Young et al.[6] made use of P-type silicon originally doped with aluminum, gallium, and indium and determined the activation energy of the shallow acceptors after neutron irradiation and annealing. The relevant data on the shallow centers are shown in Table III. The activation energies of shallow defect centers in neutron-irradiated silicon after annealing are generally smaller than the activation energies of Group III and V shallow impurities. These are interesting properties of these defect centers that warrant further investigation.

Table III. Shallow Defect Centers in Neutron-Irradiated Silicon.

Original dopant	Annealing condition	Shallow center	Activation energy eV		Ref.
			E_1	E_2	
Al	600° 1 hr.	acceptor	0.030	0.041	6
In	600° 1 hr.	acceptor	0.031	0.045	6
Ga	575-625°C 1 hr.	acceptor	0.027	0.039	6
	850°C 1 hr.	acceptor		0.074	
H	450°C 0.5 hr.	donor	0.026		This
	500°C 0.5 hr.	donor	0.026	0.037	work

Figure 4 shows a comparison between the annealing curves of samples B and D and shows that sample D with preannealing at 650°C remains P-type during annealing in the range 350-600°C because the Si-H bond in it had been previously broken (the corresponding Si-H absorption peak in Fig. 5 entirely vanishes). Although the hydrogen concentration in samples D and B is basically the same, the annealing behavior of sample D is different from that of B, but similar to that of NTD silicon grown in argon. This implies that the formation of this new donor is not simply related to the hydrogen content in silicon, but related to the existence of Si-H bonds, and thus dependent on the details of the total past thermal history of the material.

b. Optical characteristics of the new donor

The spectra of Fig. 5 show that the absorption intensity of the peak of sample A at 2150 cm^{-1} abruptly increases after annealing at 350°C, reaches its maximum after annealing at 400°C, decreases abruptly after annealing at 500°C, and vanishes after annealing at 550°C. This correlates to the changes of the donor concentration in the material shown in Fig. 4. Similar results are obtained for sample B. These considerations lead to the prediction that the peak at 2150 cm^{-1} in infrared absorption spectra may have a connection with the hydrogen-related new donor. Taking this peak as an "indicator" for the existence of the new donor helps to explain the observed annealing behavior.

The sample C without neutron-irradiation still has a peak at 2150 cm^{-1} with very weak intensity after the annealing at 400°C. Its resistivity correspondingly decreases to minimum (Fig. 4). When the annealing temperature increases to 500-550°C, the peak at 2150 cm^{-1} disappears and its resistivity returns to the original value. It is reasonable to generalize the results from samples A and B and say that this also resulted from the hydrogen-related

donor. On the other hand, a peak at 2150 cm^{-1} also exists, though
very weak, in irradiated but unannealed samples A and B. This
indicates that the hydrogen-related donor already exists after
irradiation, but with a very low concentration insufficient to
influence the net conductivity type. It can be concluded that the
existence of this new donor has a connection not only with Si-H
bond, but also with some irradiation-induced defect structures in
silicon. The concentration of these defects reaches its maximum
after annealing at about 400°C, and neutron irradiation leads to
an increase of these defects in irradiated silicon, significantly
enhancing the hydrogen-related donor effect. Further study is
needed to more fully understand the structure of such defects.

(3) Hydrogen-related deep centers

    DLTS measurements (Table II) show that samples annealed at
400°C (sample A) or at 450°C (sample B) exhibit at least six deep
centers. Figure 7 shows the positions of deep-level centers in
NTD silicon grown in argon (Guldberg[7]), in vacuum (Baliga et
al.[8]), and in hydrogen (this work). From comparison, it can be

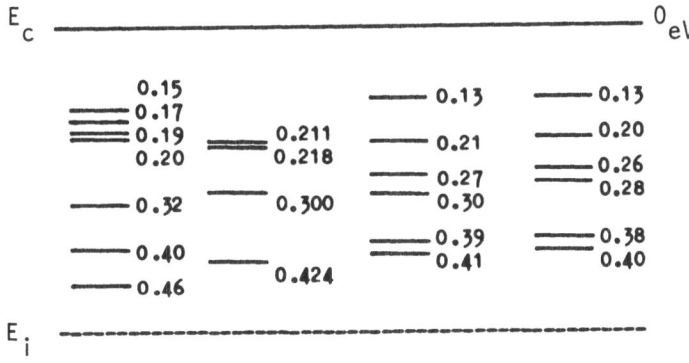

Author	Guldberg	Baliga	This Paper	This Paper
Growth Cond.	FZ in Ar	CZ in Vac.	FZ in H_2(A)	FZ in H_2(B)
Slow/Fast Flux Ratio	50	250	100	5
$\phi_{th}t(n\ cm^{-2})$	8×10^{17}	3.5×10^{17}	9.7×10^{16}	1.8×10^{17}
Anneal Temperature	450°C	455°C	400°C	450°C
Anneal Temperature	30 min	15 min	30 min	30 min

Fig. 7. Position of the deep level centers, electron traps in NTD
in silicon.

Table IV.   Variations of Resistivity ($\Omega \cdot$cm) After a Thermal
            Treatment Cycle in Samples A and B.

Sample	First anneal 30 min	Second anneal 1250°C, 30 min	Third anneal 450°C, 30 min
A	15.9 (400°C)	217	216
B	14.4 (450°C)	141	138

seen that the electron trap with $\Delta E = (0.26 - 0.27)$ eV is possibly related to hydrogen in silicon.

(4) A comparison of the hydrogen-related new donor with the oxygen
    donor formed in CZ silicon after annealing at 450°C

    a. The new donor does not exhibit the reversibility with temperature cycling of the thermal treatment (see Table IV) as does the oxygen-related donor.

    b. When the sample is annealed for 0.5 hr. or for 3 hrs. at 450°C, there is no noticeable change of the resistivity. Apparently, the concentration of the new donor does not increase with the increase of annealing time at 450°C as the oxygen donor does.

    c. The oxygen content in our original FZ-Si is two orders of magnitude lower than the oxygen content of CZ-Si and one would not expect to see oxygen donor formation in this material.

    d. The measured energy levels of the new hydrogen-related donors are different from those of the oxygen donor formed after annealing at 450°C.

    Therefore, it can be concluded that the new donors formed in NTD silicon FZ grown in hydrogen are not related to the oxygen donor.

(5) A discussion concerning the hydrogen doping method

    The samples for the above studies on the behavior of the hydrogen-related donor in silicon are generally prepared by means of hydrogen ion implantation (proton bombardment), hydrogenized amorphous silicon film growth, etc. In these methods, hydrogen concentrates in a very thin layer (~1 micron) near the surface. Kleinhenz et al.[10] made a comparison of the results of the infrared measurements in the case of hydrogen implantation with various concentrations. He could only distinguish a signal when the concentration of hydrogen was greater than 10^{20} cm^{-3} and even then he could not use it to study the physical state of the hydrogen in his silicon specimens.

In NTD silicon grown in hydrogen, the hydrogen content was estimated to be 10^{16}-10^{17} cm^{-3}, four orders of magnitude lower than in Kleinhenz's samples. The defect distribution formed in bulk single-crystal silicon is comparatively homogeneous, due to a homogeneous distribution of hydrogen and the large neutron penetration during NTD doping. Therefore, this sort of material is very suitable for the study of electrical and optical properties of hydrogen in silicon. Furthermore, the sample processing and measurement, etc. on thick bulk silicon are more convenient than in thin films. Therefore, the present results do not contradict Kleinhenz's results but extend the measurements to lower hydrogen densities and utilize a wider variety of characterization techniques.

5. CONCLUSION

(1) Neutron transmutation-doped silicon FZ grown in hydrogen exhibits an interesting behavior during annealing in the range of 350-600°C. This behavior is attributed to the formation of a hydrogen-related donor.

(2) We conclude that this new hydrogen-related donor is associated with a hydrogen complex. Its formation depends not only on the existence of the Si-H bond, but also on the defects with a special structure and their changes during elevated temperature processing. After annealing at 400°C, the concentration of this new donor reaches its maximum, the activation energy is (26 ± 1) meV after annealing at 450°C and (26 ± 1) meV and (37 ± 1) meV after annealing at 500°C. The new donor disassociates and vanishes after annealing above 550°C. In these annealing processes, the changes in concentration of this donor center correlate to the changes in intensity of the infrared absorption peak at 2150 cm^{-1}.

(3) This hydrogen-related new donor is distinctly different from the well-known oxygen-related donor formed in CZ silicon during annealing at 450°C.

(4) After annealing at about 400°C, NTD silicon FZ grown in hydrogen provides a very suitable vehicle for studies of the behavior of hydrogen in silicon.

(5) A thorough theoretical analysis and a comprehensive experimental study of the structure and electronic configuration of the hydrogen-related new donor is necessary to fully understand and interpret these experimental observations.

ACKNOWLEDGMENT

    We are grateful to the Shanghai Electronic Devices Material
Works and to the Loyang Single Crystal Silicon Factory for supply-
ing the silicon specimens used in the present experiments and to
The Atomic Energy Institute, Academia Sinica, and the Nuclear
Energy Institute, Tsinghua University for performing the NTD dop-
ing on these specimens. We wish to acknowledge the help of
Messrs. Wang Wannian, Shan Lei, Feng Yi, Zhou Yuan, Ruan Shenyan,
Xu Xuemin and Lian Junwu in the experiments and analysis of exper-
imental results and Mr. Zhang Zhengnan for measurements of infrared absorption spectra.

REFERENCES

1. Y. Zohta, Y. Ohmura, and M. Kanazawa: Japan J. Appl. Phys.,
    10,532 (1971)

2. S. T. Picraux, F. V. Vook, and H. J. Stein: "Conf. Ser. No.
    46. IOP. Defects and Radiation Effects in Semiconductors,"
    31. (1978)

3. H. J. Stein: J. Electronic Mat., 4, 159 (1975).

4. V. A. Singh, C. Weigel, J. W. Corbett, and L. M. Roth: Phys.
    Stat. Sol., (B) 81, 637 (1977).

5. V. A. Singh, J. W. Corbett, C. Weigel, and L. M. Roth: Phys.
    Letters., 65 A, 261 (1978).

6. M. H. Young, O. J. Marsh, and R. Baron: "Proc. 2nd Intern.
    Conf. on Neutron Transmutation Doping in Semiconductors,"
    335 (1979).

7. J. Guldberg: J. Phys. D: Appl. Phys., 11, 2043 (1978).

8. B. J. Baliga and A. D. Evwaraye: "Proc. 2nd Intern. Conf. on
    Neutron Transmutation Doping in Semiconductors," 317 (1979).

9. P. Gaworzewski and K. Schmalz: Phys. Stat. Sol., (A) 55,
    699 (1979).

10. R. L. Kleinhenz, Y. H. Lee, V. A. Singh, P. M. Mooney, A.
    Jaworowski, J. C. Corelli, and J. W. Corbett: "Conf. Ser.
    No. 46. IOP. Defects and Radiation Effects in Semiconduc-
    tors," 200 (1978).

PARTICIPANTS

Afsar, M.N.
Francis Bitter National Magnet Laboratory, 170 Albany St., Massachusetts Institute of Technology, Cambridge, MA   02139

Akhmetov, V.D.
Institute of Semiconductor Physics, Novosibirsk-90, USSR

Alm, A.O.
ASEA AB, Fack, Electronics Division, Semiconductor Components Department S-721 83 Vasteras, Sweden

Badham, K.A.
Atomic Energy of Canada Limited, Commercial Products, P.O. Box 6300, Ottawa, Ontario, Canada   K2A 3W3

Baliga, B.J.
General Electric Company, Corporate Research & Development Center, Schenectady, NY   12345

Birtcher, R.C.
Argonne National Laboratory, Materials Science Division, 9700 S. Cass Ave., Argonne, IL   60439

Blackburn, D.L.
National Bureau of Standards, Bldg. 225, Rm. B310, Washington, DC   20234

Breant, P.
Commissariat a l'Energie Atomique, Rue de la Federation 35, Paris Cedex XV, France 75

Cao, Y.Z.
Institute of Atomic Energy, P.O. Box 275, Beijing, People's Republic of China

Carter, R.S.
National Bureau of Standards, Bldg. 235, Rm. A106, Washington, DC   20234

327

Chandrasekhar, H.R.   Research Reactor Facility, University of
                          Missouri, Columbia, MO 65211

Chu, E.K.   Westinghouse Electric Corporation,
                          Semiconductor Division, MS 20, Youngwood,
                          PA 15697

Covington, B.   Sam Houston State University, Physics
                          Dept., Huntsville, TX 77341

Crees, D.E.   Marconi Electronic Devices Ltd.,
                          Carholme Road, Lincoln, LN1 1SG, England

Farmer, J.W.   Research Reactor Facility, University of
                          Missouri, Columbia, MO 65211

Fecych, W.   Massachusetts Institute of Technology,
                          138 Albany St., Cambridge, MA 02139

Gao, J.   Institute of Atomic Energy, P.O. Box
                          275, Beijing, People's Republic of
                          China

Gunn, S.   Research Reactor Facility, University of
                          Missouri, Columbia, MO 65211

Haller, E.E.   Lawrence Berkeley Laboratory, One
                          Cyclotron Road, University of
                          California, Berkeley, CA 94720

Hansen, K.   Risø National Laboratory, Post Box 49,
                          DK 4000 Roskilde, Denmark

Hart, R.R.   Texas A & M University, Department of
                          Nuclear Engineering, College Station, TX
                          77843

Henry, K.J.   United Kingdom Atomic Energy Authority
                          Harwell, Didcot, Oxfordshire, OX11 ORA,
                          England

Herzer, H.   Wacker-Chemitronic, Burghausen, West
                          Germany

Heydorn, K.   Isotope Division, Risø National
                          Laboratory, DK 4000 Roskilde, Denmark

Kaltenborn, N.A.   Institute for Energy Technology, P.O.
                          Box 40, N-2007 Kjeller, Norway

Kobisk, E.H.	Oak Ridge National Laboratory, P.O. Box X, Oak Ridge, TN 37830
Kramer, H.G.	Monsanto Company, P.O. Box 8, St. Peters, MO 63376
Larrabee, R.D.	National Bureau of Standards, Bldg. 225, Rm. A331, Washington, DC 20234
Lu, C.	Institute of Atomic Energy, Beijing, People's Republic of China
Malmros, O.	Topsil A/S, Linderupvej 4, DK 3600 Frederikssund, Denmark
Marsh, O.J.	Hughes Research Laboratories, 3011 Malibu Canyon Road, Malibu, CA 90265
Meese, J.M.	Research Reactor Facility, University of Missouri, Columbia, MO 65211
Mitchel, W.	Air Force Materials Laboratory, AFWAL/MLPO, Wright Patterson AFB, OH 45433
Murphy, T.	Atomic Energy of Canada Limited, Commercial Products, P.O. Box 6300, Ottawa, Ontario, Canada K2A 3W3
Nicholson, G.	Atomic Energy of Canada Limited, Chalk River Nuclear Laboratories, Chalk River, Ontario, Canada K0J 1P0
Nielsen, K.F.	Risø National Laboratory, Frederiksborgvej, 4000 Roskilde, Denmark
Pajot, B.	Groupe de Physique des Solides de l'Ecole Normale Superieure, University of Paris VII, 75251 Paris Cedex 05, France
Paxman, D.H.	Philips Research Laboratories, Cross Dark Lane, Salforas, Redhill, RH1 5HA, Surrey, England
Rundquist, H.	Studsvik Energiteknik AB, S-61182 Nykoping, Sweden
Ruzicka, W.G.	Union Carbide, P.O. Box 324, Longmeadow Road, Tuxedo, NY 10987

Sittig, R.   Brown, Boveri & Cie, Department EKS, Werk
                              Birr, CH 5242 Birr, Switzerland

Smith, T.G.G.   United Kingdom Atomic Energy Authority,
                              Didcot, Oxfordshire, England OX11 ARA

Stone, B.D.   Monsanto Company, P.O. Box 8, St. Peters,
                              MO 63376

Vieweg-Gutberlet, F.G.   Wacker-Chemitronic GMBH, Technical
                              Marketing, P.O. Box 1140, 8263 Burghausen,
                              West Germany

Wang, Z.   Institute of Semiconductors, Acadamia
                              Sinica, Beijing, People's Republic of
                              China

Ward, C.J.   Monsanto Company, 755 Page Mill Road, Palo
                              Alto, CA 94304

Watson, J.B.   Howard Payne University, HPU Station,
                              Brownwood, TX 76801

Wolverton, W.M.   Monsanto Company, P.O. Box 8, St. Peters,
                              MO 63376

Young, M.H.   Hughes Research Laboratories, 3011 Malibu
                              Canyon Road, Malibu, CA 90265

Zhang, Z.   Institute of Atomic Energy, P.O. Box 275,
                              Beijing, People's Republic of China

INDEX*

Annealing of NTD Damage
  during device fabrication, 305-309
  in cadmium sulfide, 70-74
  in cadmium telluride, 70-74
  in gallium arsenide, 16-19, 74-75
  in silicon, 67-70, 171-175, 194-198, 206
  - characterization by DLTS, 75-79
  - characterization by infrared absorption, 297-303
  - characterization by Raman scattering, 287-294
  - grown in hydrogen atmosphere, 311-325
Annual NTD-silicon production, 86, 89, 158-161, 163
Argonne intense pulsed neutron source, 54-57
Axial resistivity gradient, see resistivity, uniformity of

Bolometers, semiconductor
  advantages of NTD material, 27, 35
  conduction mechanisms in, 26-27
  NTD germanium, properties of, 30-35
  theory of, 22-26

Cadmium sulfide
  irradiation damage and conductivity, 58, 60-62
  isochronal annealing, 70-74
Cadmium telluride
  irradiation damage and conductivity, 58, 60-62
  isochronal annealing, 70-74
Compensation
  in indium-doped silicon, 261-269
  - importance of, 261-262
  - infrared absorption, 265-269
  measured by photoluminescence, 241, 243-246
Deep-level transient spectroscopy (DLTS)
  comparison to TCS, 225-226, 238

---

*Unless otherwise indicated, the items in this index refer to
 silicon.